Dear Professor -
Thank you for being
a true inspiration +
setting the bar high.
You will have a lasting
imprint at SBCC.
— Colby

Dear Al,

The department will not
be the same without you,
neither will be the college.
We will miss you so
much!

Cornelia

Dear Al,
This has been
a true pleasure
working with you! -
Il miss your wit
and good thoughts.

Barbara Anne

Maps and the 20th Century

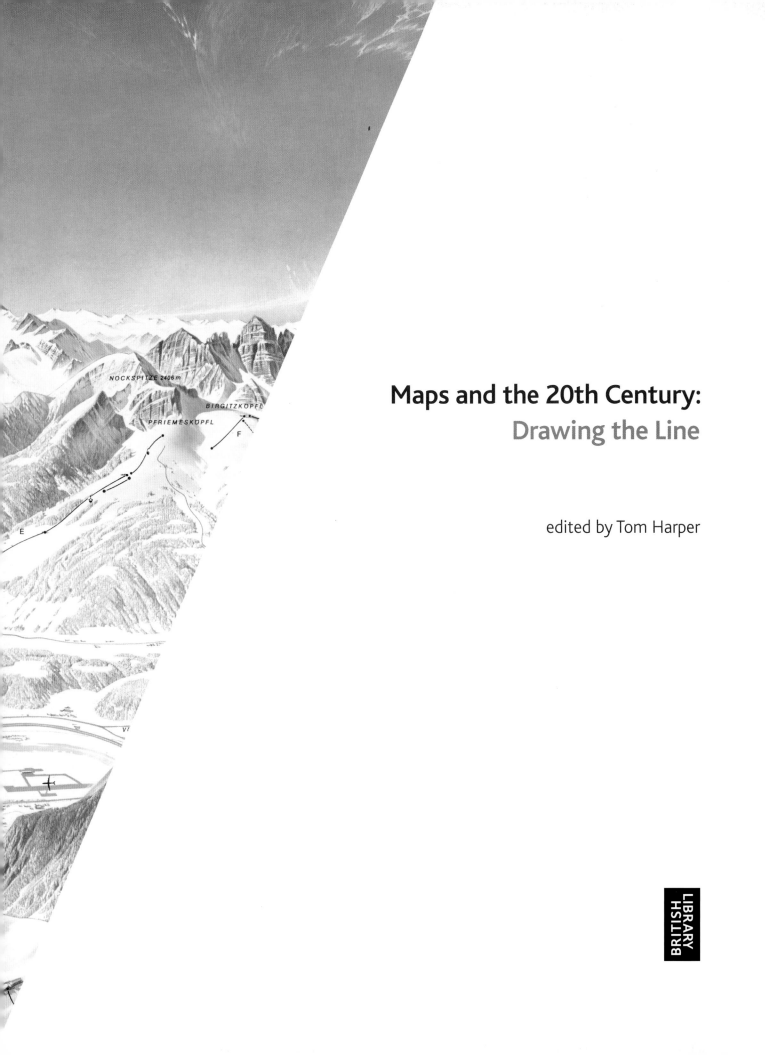

Maps and the 20th Century:
Drawing the Line

edited by Tom Harper

First published in 2016 by
The British Library
96 Euston Road
London NW1 2DB

On the occasion of the British Library exhibition
Maps and the 20th Century: Drawing the Line
4 November 2016 – 1 March 2017

British Library Cataloguing in Publication Data
A catalogue record for this book is available from
the British Library

ISBN 978 0 7123 5661 9 (paperback)
ISBN 978 0 7123 5662 6 (hardback)

Designed by Briony Hartley, Goldust Design
Picture research by Sally Nicholls
Printed in Hong Kong by Great Wall Printing Co. Ltd

CONTENTS

INTRODUCTION
Drawing the Line

Tom Harper

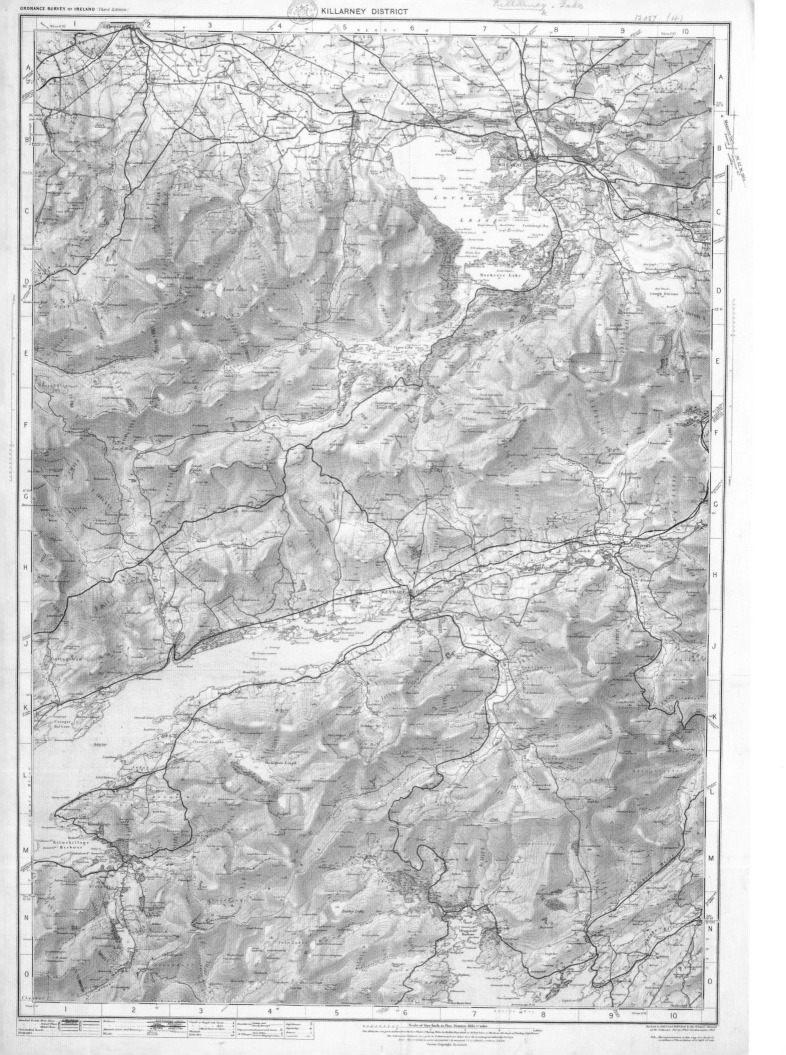

We do recognise, of course, that our subjective experience can take us into realms of perception, imagination, fiction, and fantasy, which produce mental spaces and maps as so many mirages of the supposedly 'real' thing
David Harvey, *The Condition of Postmodernity*, 1989

This is where I am just now. I'm still 'on the map' you see!
Postcard by Arthur John Carter, *c.* 1914

Maps were active agents of change in the twentieth century, and how they affected the course of history is the subject of this book.[1] This introduction sets out the basis for the cartographic century. As objects, maps became accessible and appreciable to mass society for the first time, and enabled people to place themselves 'on the map'. As images, even more than in previous eras, they were understood by people as objective replications of reality, blurring and even dissolving the difference in their viewers' minds between maps and what they purported to represent (ill. 1). Thanks to both of these developments, maps were able to act as even more powerful cultural mediators between people and their political, economic and social masters, reinforcing existing realities and creating new ones.

There are many different maps in the world, showing different things in different ways for different reasons. They cannot all be correct, nor can they ever convey the complete and continuous totality of a place. The important point to make is that every single map ever made presents a version of the world, not the real world itself; the 'real' world we experience is made up of numerous social constructs such as those shown by maps. This understanding was only properly articulated in the twentieth century,[2] though the appreciation of maps as anything other than mirrors of reality proved difficult to shift. By convincing people of the objectivity of what they showed (and denying the existence of what they left out), maps changed the world.

In order to understand the nature of this change, it is necessary to appreciate the basis for the cartographic century as the period of European world dominance, which began in the sixteenth century and, passing through the period of Enlightenment, discovery and advances in science, reached its apogee in the industrialised imperial nineteenth century. During that time in Europe and North America mass education and technological changes in map production brought cartography to a wide audience – not just, as previously, the privileged elite.

After the high-water mark of this European imperialist project in 1914, momentous changes were witnessed through maps: in war, where maps enabled conflict to be waged

1 (left)
Killarney District. Ordnance Survey, 1913. The Ordnance Survey's maps were regarded as unimpeachable replications of reality.
Maps 12087.(4)

2 (next page)
Alighiero e Boetti, *Mappa*, Rome, 1990-91. Alighiero Boetti's embroidered maps emphasised the clusters of post-colonial states in Africa, for example

quickly and comprehensively, though not necessarily more clinically; in peace, where fresh border lines drawn on maps dictated the futures of whole populations (ill. 2); in economics, where maps perpetuated and protected financial structures and commercial practices, becoming commodities in their own right; and in movement and change, capturing and propelling the dynamic and turbulent essence of the twentieth century forward by illustrating travel, migration, ocean currents and plate tectonics, for example. Everyday life was changed irrevocably by these things, and the chapters in this book will explore each of them more closely. In preparation, here I will outline how maps came to dominate twentieth-century society's understanding of the world, and how they dealt with the overwhelming changes encountered.

Finding the way onto the map

Imagine a time when a map would have been seen as a novelty. For a large number of people in Western society this was still just about the case at the beginning of the twentieth century. Compulsory education (in Britain from 1870, but substantially earlier on much of the Continent) made them more familiar with maps. When learning about geography at school at the beginning of the century, pupils were taught to develop their geographical imagination by viewing the world as if they were placed upon a map, and projecting their viewpoint from inside it (ill. 3). From this point, thanks to the vastly increasing number of cartographic postcards, news maps and other cheap map products, finding and marking oneself and others' positions with a dot or cross became a natural thing to do. The novelty was still visible in the map postcard of circa 1914 sent by a holidaymaker in the faraway British Lake District (ill. 4).

3 (left)
Boys at work in a British classroom, 1922.
Location unknown.

4 (right)
This is where I am just now, I'm still
'on the map' you see!
Postcard by John Arthur Carter, c. 1914.
Maps C.1.a.9.(199)

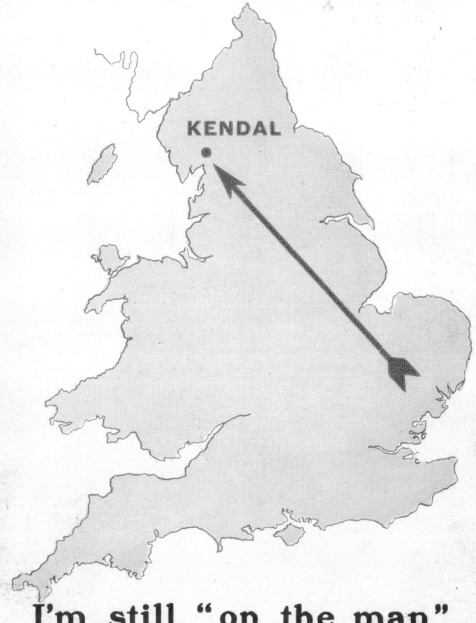

A.J.C.—268

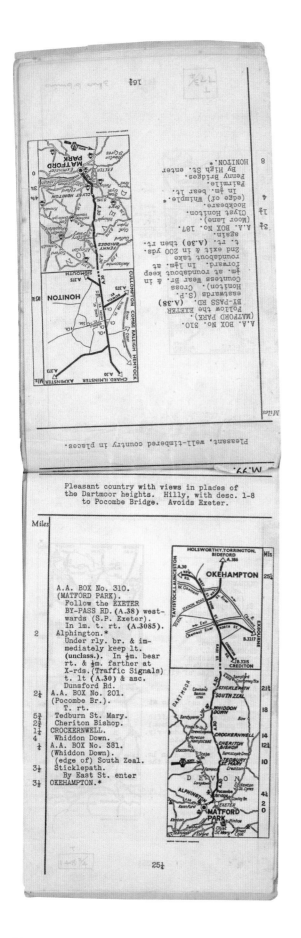

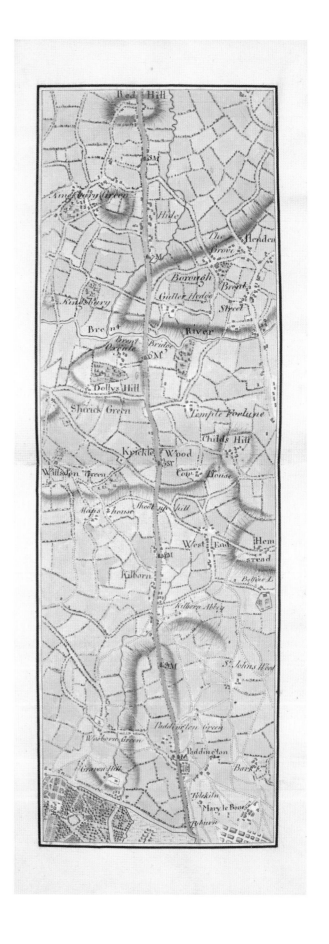

5 (far left)
Page from a personalised route guide produced by the Automobile Association for one of its members, showing a section of the route from Gosport to Liskeard. c.1965
Maps CC.5.b.50.(1)

6 (left)
A 'ribbon map' of the road from London to Luton Park, prepared for John Stuart, 3rd Earl of Bute, after maps by John Rocque, 1767.
Add.MS 74215

7 (below)
The story of a group of students who set sail to find the world's most remote island was featured in the *Evening Express*, Thursday, 1 September 1955

However, the novelty would wear off. Increasing personalisation of maps and tailor-made products placed the user more routinely at their centre. In previous centuries this privilege had been the almost exclusive preserve of royalty. But by 1911 the Automobile Association of the UK was compiling personalised map itineraries for its thousands of members (ills 5, 6). This is the genesis of the twenty-first-century blue dot, representing the map user at the centre of his or her personalised interactive map.

Existing on the map made it more difficult for people to get lost than before. So much so in fact, that by the second half of the twentieth century to 'get lost', to 'escape' or to go 'off the map' became a rare, luxurious pleasure for those who could afford it. Meanwhile, others took it as a challenge (ill. 7). There were other consequences, though. By moving onto the map, society itself became more comprehensively mapped by the military, government and commercial agencies that maintained national mapping systems, including satellites and geographic information systems (GIS). In previous centuries intrusion into people's private spaces by official surveyors had been met with angry opposition, with dogs and even weapons used against them. A factor in the Ordnance Survey's 'Battle of the Scales' of the 1840s had been concern over privacy. By the late twentieth century, however, mapping had become so comprehensive that it was difficult for people to escape from it even if they wanted to.

Thursday Sept. 1 1955, EVENING EXPRESS — 3

THEY'LL BE WORLD'S LONELIEST MEN

"**I** SAY," said a tall young man with a pipe, "I've got a marvellous recipe for fried seals' brains. First you soak them in sea-water for three hours—that makes them hard—then you slice them thin and fry them. We'll have lots of that."

This remarkable conversation took place in a top-floor flat in a house off Gloucester Road, London. There were four young men in the

By BERNARD LEVIN

large untidy room when I arrived, books and files were strewn about the floor, a tape-recorder was shrilling piano-music.

"Oh, good," said another tall young man wearing spectacles.

It looked like the living-room of a house shared by a group of undergraduates and indeed the four men had only just finished university.

But they were in the last stages of planning an enterprise beside which the most extravagant steeple-climbing ever indulged in among the dreaming spires of Oxford or Cambridge would appear rather less exciting than marbles.

They are off to spend five months on an uninhabited and largely unexplored island in the sub-Antarctic.

You've heard of Tristan Da Cunha, "the loneliest island in the world?" Well Gough Island is 260 miles from Tristan Da Cunha and just that much lonelier. It is 1500 miles from the

leading the expedition in place of John Heaney, retired.

There would be Michael Swales (he was the one who had found the recipe for seals' brains) who is one of the party's two zoologists, Philip Mullock who will be the radio operator. Chambers himself,

leaving behind them. They are taking a few gramophone records — Vaughan Williams' Antarctic Symphony for instance — which they will probably be too busy to play, and a few books including the Pilgrim's Progress, Dante's

Lonely Gough Island — a climate of fog, rain and storms.

Purgatory and Lewis Carroll's Hunting of the Snark—which they will probably be too busy to read.

THE MAP THAT

CAME TO LIFE

Longlands Wood

Tumuli

† East Stanton

Abbey

Redroof Fᵐ

8
Cover of *The Map that Came to Life*, by Henry Deverson and Ronald Lampitt. Oxford University Press, 1948.
Cup.1246.aa.53

Dissolving the gap between the map and the reality

In placing themselves on maps, people made them their surroundings. The importance of 'accuracy' in maps – accuracy being the degree to which a map replicated the dimensions and appearance of the places it represented – was an overriding feature of Enlightenment mapping. We can detect this in the titles of 'New and Accurate' eighteenth-century maps. But again in the twentieth century, thanks to teaching practices and technological change, the gap between maps and what they represented narrowed dramatically in people's perception: a conceptual leap, which was normalised through familiarity. We can see it in educational books such as Henry Deverson and Ronald Lampitt's *The Map that Came to Life* of 1948, in which two children take a journey through the landscape using an Ordnance Survey map, matching its signs and symbols to what they actually see. On the front cover of the book the foreground scene blends imperceptibly into the map (ill. 8).

Belief that the map was identical to what it purported to show was reinforced, first by the ubiquity of maps in everyday life, in all previously mentioned formats plus film – and, from the 1950s, television. Second, from the 1940s and late 1960s respectively, the movement of aerial and satellite photographic imaging ('mapping from above') into the public consciousness gave the impression that an exact, faithful and comprehensive surrogate of the world existed. Corrective adjustments to the image such as colour

and image sharpening only served to improve the appearance of the authentic reality. As a result of this layering of maps and reality, by the 1990s it was entirely normal for air passengers circling London to hum the theme tune of the BBC television drama *EastEnders* while looking out of the window (ill. 9).

The side-effects of this dissolving divide between map and reality were serious. First, it meant that any cartographic practices that did not conform to the Western objective style of mapping could not be tolerated. One of the biggest legacies of the imperial era was the unwitting obliteration of the indigenous mapping practices of cultures with which European empires had contact, because they did not match the European maps' accurate, scientific – and therefore perceived as useful – qualities. As a result, non-Western mapping traditions were often eradicated, leaving only the occasional artefact thought curious and worthy of retention (ill. 10).

The second side-effect was that since maps were held to embody the reality of what they showed, when they did show errors or something contrary to the viewer's belief, disorienting effects were pronounced. Many obvious map mistakes were and still are sources of mirth or derision. But maps showing an incorrect depiction of the extent of Arctic sea ice for example, the absence of a town, or an incorrectly delineated national border, had serious effects and dangerous consequences. If the map was the reality, then cartographic land-grabs had the potential to cause grave offence (ill. 11). Such instances were particularly common in the twentieth century, where more national borders were

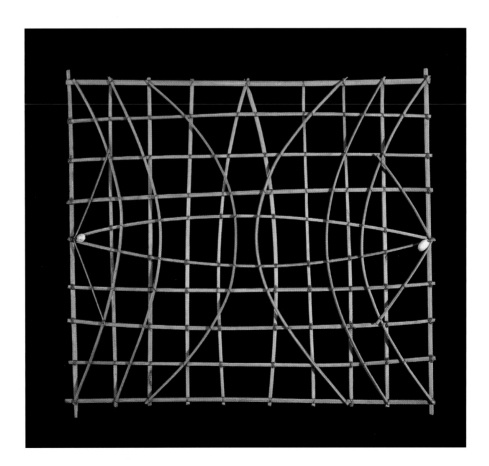

9 (left)
Screenshot of the opening credits
of the BBC television drama
EastEnders, first broadcast in 1985

10 (right)
Navigational stick chart,
produced in the Marshall Islands
c. 1900.
British Museum, Oc1941,01.4

11
Unofficial postcard showing a satellite
view of Israel and parts of Egypt, Saudi
Arabia and Jordan, labelled 'Israel'.
Published in Israel by Palphot Ltd.,
late 1970s
Maps C.12.c.1.(1824)

drawn as a consequence of war, decolonisation and the principles of national self-determination after 1918. States that refused to recognise the existence of other states excluded them from their maps (either by appropriating them within their borders, or by leaving them greyed out and unknown). Different maps created multiple realities, but variations still did not compromise the conception of the map as a mirror of nature.

Effective cultural mediators

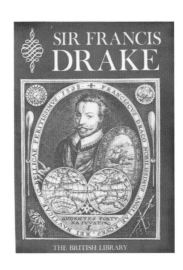

12

Sir Francis Drake: An Exhibition to Commemorate Francis Drake's Voyage Around the World, 1577-1580. British Museum Publications Ltd. for the British Library, 1977. Maps Ref.A.6a.(28)

With people anchored 'on the map', which had come to be seen as no different from the 'real' world, maps were able to work more effectively as tools of governmental, military and commercial interests, with these powers controlling the instruments of mapping as they had done for centuries. Stuart Elden has explained the political definition of 'territory' as comprising the measure and control of space.[3] In this context it is easy to appreciate the powerful mandate that state mapping had in administration and control, particularly given technological support such as GIS, which consolidated that control.

Much governmental mapping was restricted from public view. But one of the lessons learnt by authorities in the twentieth century, particularly as a consequence of the Russian Revolution of 1917, was that an enfranchised mass society could not, as before, simply be exploited or left to its own devices. Some form of mediation was needed between society and its masters, and maps became one of the devices with which this relationship was more strongly maintained. Maps functioned in a number of ways. As objects of leisure they accompanied activities such as travelling or national celebration, particularly following periods of war; happy people were less likely to revolt. Maps continued to be used symbolically, with the recognisable shapes of states and countries – particularly newly independent ones – appearing on coins and emblems as symbols of loyalty and identity.

Older maps had played this latter role for longer. During the mid-nineteenth century the United States Hydrographic Office and Congressional Library were encouraged by the German geographer Johann Georg Kohl to build a collection of old maps of America, because possession of such maps reinforced the validity of the still young nation's existence. The purchasing of cartographic heirlooms by national libraries continued throughout the twentieth century, and beyond. They were used in exhibitions celebrating the anniversaries of famous explorers such as Columbus, Drake and Cook – national heroes and symbols of national achievement (ill. 12). Production of the vast facsimile atlas of historical Portuguese charts, the *Portugaliae Monumenta Cartographica* of 1962, was bankrolled by the authoritarian regime of António de Oliveira Salazar as an exercise in legitimacy and to support its policy towards its crumbling colonial empire.

Historical maps mediated in a different way from modern cartography, which after 1950 was used to demonstrate key ideological concepts such as progress, improvement and development. The political picture of fast-moving and positive national change, aided by science, was promoted through national atlases and the scientifically based mapping used in planning, administration and welfare.

The cartographic alter ego of scientific mapping was propaganda mapping. Overt, powerful and persuasive cartography became a tool used by governments to influence and control their populations more strongly during the twentieth century. It became highly developed particularly during periods of conflict, and as a consequence the twentieth century was littered with images that used maps to distort and manipulate reality, and to produce emotions such as fear, anger or triumph, and above all loyalty to the nation or cause (ill. 13). For example, maps were drawn with distortion or exaggeration in order to incite distress or outrage. Other techniques, such as the use of warm colours and the inclusion of happy, smiling faces, were employed during peacetime (ill. 14).

These are some of the ways in which maps mediated between mass society and its masters: a form of soft control in which people, from their vantage point on the map-come-to-life, gained a sense of identity and belonging. Though it builds on the local and national patriotism of earlier periods, the strong empathy between mass society and physical places is a phenomenon of the twentieth century.

13 (left)
Kärnten in Gefahr!
Vienna, *c.* 1919. An Austrian poster protesting at the potential loss of the region of Carinthia to Yugoslavia (symbolised by the threatening red claw).
Maps CC.6.a.82

14 (right)
J. P. Sayer, *The Map of Peace / Keep on Saving, We've Great Things to Do.*
National Savings Committee, 1945.
Maps CC.5.b.51

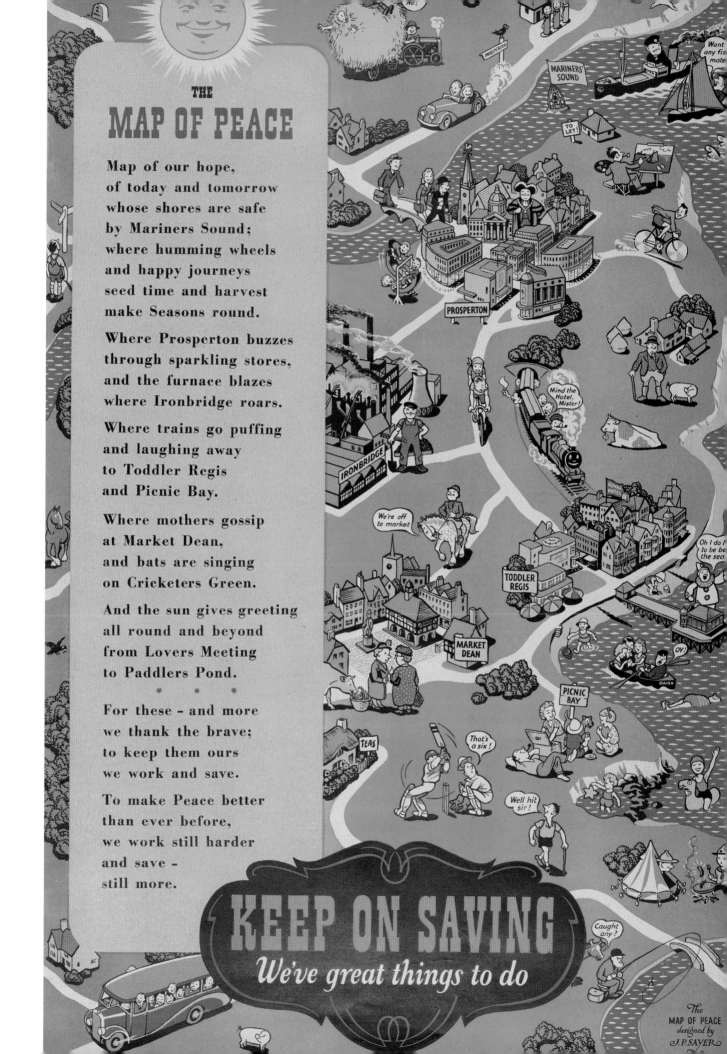

THE MAP OF PEACE

Map of our hope,
of today and tomorrow
whose shores are safe
by Mariners Sound;
where humming wheels
and happy journeys
seed time and harvest
make Seasons round.

Where Prosperton buzzes
through sparkling stores,
and the furnace blazes
where Ironbridge roars.

Where trains go puffing
and laughing away
to Toddler Regis
and Picnic Bay.

Where mothers gossip
at Market Dean,
and bats are singing
on Cricketers Green.

And the sun gives greeting
all round and beyond
from Lovers Meeting
to Paddlers Pond.

*   *   *

For these – and more
we thank the brave;
to keep them ours
we work and save.

To make Peace better
than ever before,
we work still harder
and save –
still more.

KEEP ON SAVING
We've great things to do

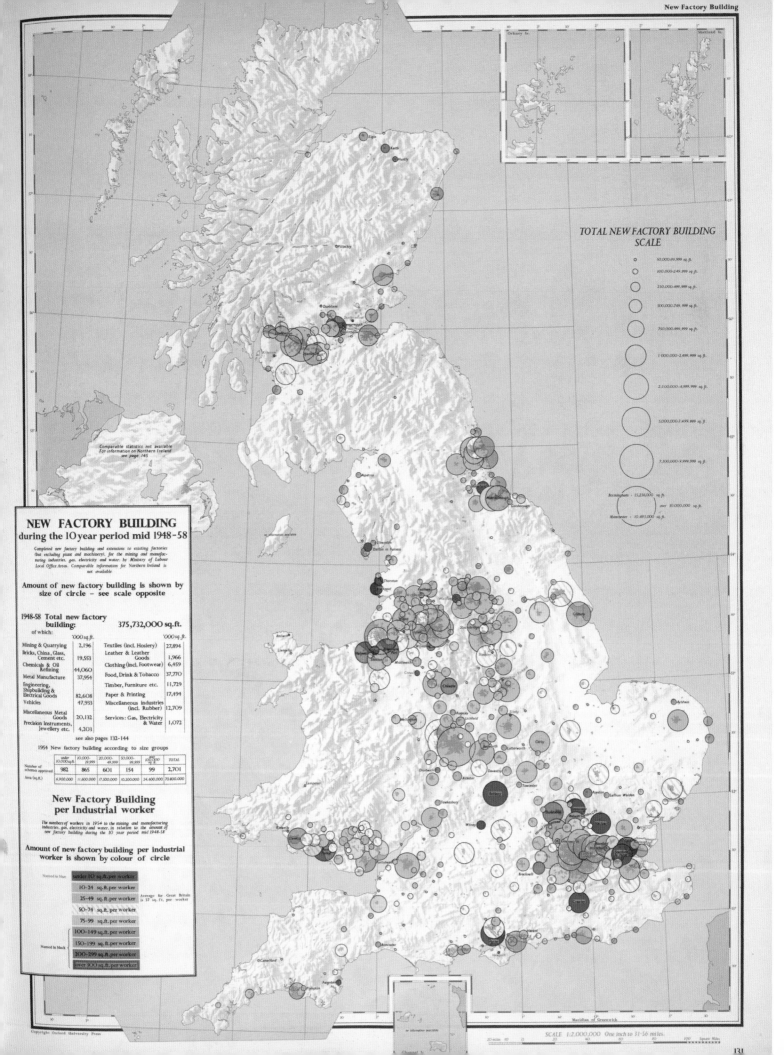

TOTAL NEW FACTORY BUILDING SCALE

- 50,000-99,999 sq.ft.
- 100,000-249,999 sq.ft.
- 250,000-499,999 sq.ft.
- 500,000-749,999 sq.ft.
- 750,000-999,999 sq.ft.
- 1,000,000-2,499,999 sq.ft.
- 2,500,000-4,999,999 sq.ft.
- 5,000,000-7,499,999 sq.ft.
- 7,500,000-9,999,999 sq.ft.

Birmingham – 15,256,000 sq.ft.
over 10,000,000 sq.ft.

Manchester – 10,493,000 sq.ft.

Comparable statistics not available
For information on Northern Ireland
see page 145

NEW FACTORY BUILDING
during the 10 year period mid 1948-58

Completed new factory building and extensions to existing factories
(but excluding plant and machinery), for the mining and manufacturing industries, gas, electricity and water, by Ministry of Labour
Local Office Areas. Comparable information for Northern Ireland is
not available

**Amount of new factory building is shown by
size of circle – see scale opposite**

1948-58 Total new factory building: 375,732,000 sq.ft.
of which:

	'000 sq.ft.		'000 sq.ft.
Mining & Quarrying	2,196	Textiles (incl. Hosiery)	27,894
Bricks, China, Glass, Cement etc.	19,553	Leather & Leather Goods	1,966
Chemicals & Oil Refining	44,060	Clothing (incl. Footwear)	6,459
Metal Manufacture	37,954	Food, Drink & Tobacco	37,770
Engineering, Shipbuilding & Electrical Goods	82,608	Timber, Furniture etc.	11,729
Vehicles	47,933	Paper & Printing	17,494
Miscellaneous Metal Goods	20,132	Miscellaneous industries (incl. Rubber)	12,709
Precision Instruments, Jewellery etc.	4,203	Services: Gas, Electricity & Water	1,072

see also pages 132-144

1954 New factory building according to size groups

	under 10,000 sq.ft.	10,000-19,999	20,000-49,999	50,000-99,999	over 100,000 sq.ft.	TOTAL
Number of schemes approved	982	865	601	154	99	2,701
Area (sq.ft.)	6,900,000	11,600,000	17,500,000	10,200,000	24,600,000	70,800,000

New Factory Building per Industrial worker

The numbers of workers in 1954 in the mining and manufacturing
industries, gas, electricity and water, in relation to the amount of
new factory building during the 10 year period mid 1948-58

**Amount of new factory building per industrial
worker is shown by colour of circle**

Named in blue	under 10 sq.ft. per worker
	10-24 sq.ft. per worker
	25-49 sq.ft. per worker — Average for Great Britain is 37 sq.ft. per worker
	50-74 sq.ft. per worker
	75-99 sq.ft. per worker
	100-149 sq.ft. per worker
	150-199 sq.ft. per worker
Named in black	300-299 sq.ft. per worker
	over 300 sq.ft. per worker

Copyright Oxford University Press

SCALE 1:2,000,000 One inch to 31·56 miles.

20 min. 10   20   40   60   80   100 Square Miles

Meridian of Greenwich

Maps become self-conscious objects

Maps became self-conscious objects after the Second World War. Geographers started asking questions of maps, such as 'How is it that maps are such convincing images in everyday life?' and 'Why do maps deliberately mislead?' The latter question arose as a reaction to the large quantities of distorted and deceitful maps that had been issued as wartime propaganda, and that had two consequences. The first was that maps exhibiting an obvious degree of coercion, including propaganda, but also advertising and other popular map forms, were categorised as 'non-maps' by many geographers, academics and appreciable portions of society.

This 'narrowing of the cartographic canon', as John Pickles has put it, was also motivated by a second consequence, again partly a reaction to wartime propaganda mapping, wherein maps became defined and developed as incorruptible, scientific conveyors of information, communication models and test environments for reality.[4] It developed in earnest in the 1960s during the scientific and technological revolution (ill. 15). Pure, honest cartography was even aligned with the democratic ideal. For practical and ideological purposes, therefore, deliberate and accidental errors in maps were not tolerated by geographers (though the blanked-out areas of sensitive military sites attest to their continued toleration by the authorities).

This is the reason why today we regard some maps – the objective scientific ones – as more 'map-like' than others. Given the centrality of maps to people's perceived worlds and identities, it is easy to appreciate the danger that doubt in maps' reliability could have caused. Indeed, most people continued to filter out maps they didn't like, want or need, and selected them to fit their reality. But the threats to this established vision, coming after 1945 and at the end of the 1960s during periods of enormous transition in global economic, social and political processes and structures, were real. For some people, maps no longer matched the world they now knew.

Mapping from below

Governments continued to use maps to mediate with society. But when by the early 1970s the post-war 'Golden Age' came to an end with economic crises (1973 onwards), insurgency and revolution (Northern Ireland from 1969; Ethiopia, 1974; Iran, 1979), the end of consensus politics (1976), protests (1968) and the loss of US legitimacy in the eyes of the Left over the Vietnam War (1955–75), the landscape no longer matched them. Emerging postmodern theories questioned fundamentally held certainties about the world such as progress, modernism, improvement – and maps.

People began to make their own maps. 'Psychogeographers' such as Guy Debord did things like cut up urban maps and rearrange them to represent journeys of emotion or chance, journeys with motives other than those of everyday necessity (ill. 16). Maps were created and used for protest in art galleries or to assist demonstrations. Maps were also created out of empathy; the image of the globe that had traditionally symbolised power

15 (left)
New Factory Building. Scientific economic map from *The Atlas of Britain.* Clarendon Press, 1963. Maps Ref C.10.(3)

16 (next page)
Pages from Guy Debord and Asger Jorn's seminal psychogeographical work *Mémoires.* Copenhagen, 1959. The artwork seen here includes fragments of a map of Paris. C.188

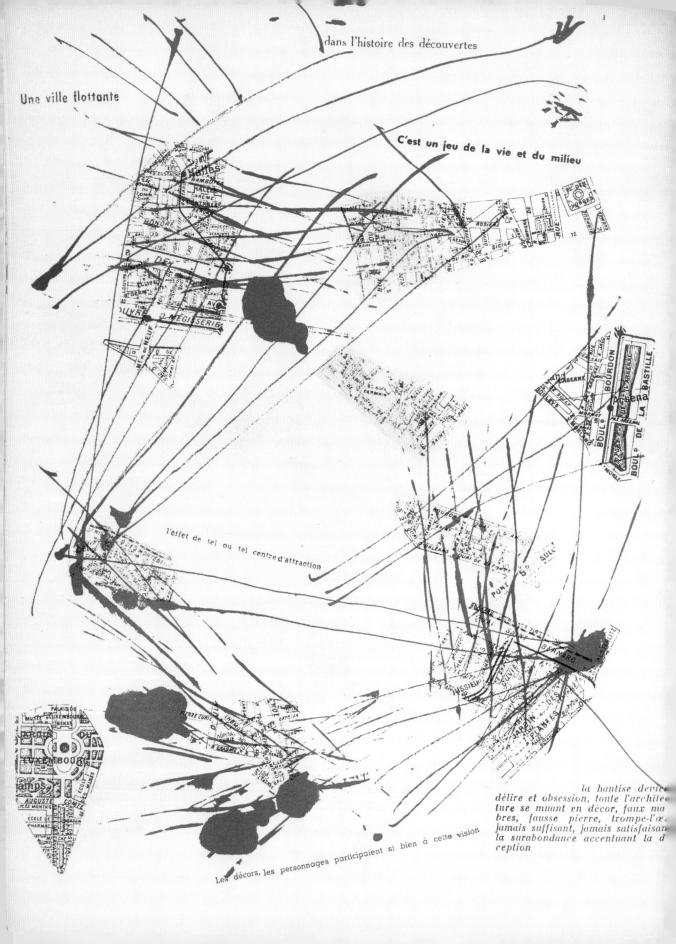

dans l'histoire des découvertes

Une ville flottante

C'est un jeu de la vie et du milieu

l'effet de tel ou tel centre d'attraction

la hantise devie[nt]
délire et obsession, toute l'architec[-]
ture se muant en décor, faux ma[r-]
bres, fausse pierre, trompe-l'œ[il]
jamais suffisant, jamais satisfaisan[t]
la surabondance accentuant la d[é-]
ception

Les décors, les personnages participaient si bien à cette vision

dans le feu des injures, des menaces, des exécrations et des blasphèmes

Il est sans doute trop tôt

un avis répandu sur les murs de Paris annonça le passage éphémère

spectacle, sans autre spécialité bien définie que le scandale

ouvert toute la nuit

and worldly knowledge was appropriated to highlight environmental concern for the fragile Earth.

'Mapping from below' was a phenomenon that would not have been possible in any earlier century. By the 1970s most people were map-literate and had the means of mapping at their disposal. But most importantly, some were motivated to drive a wedge between the map and the reality, to shine a mirror upon the injustices of the world, which had seen the gap between rich and poor widen and many old inequalities remain. Alternative realities offered representation for the traditionally non-Western, non-male, poor and dispossessed, who hitherto had always found themselves alienated – off the map.

Against the plethora of map worlds in existence by the 1990s and the profound disorientation caused by the fall of Communism (1989–91), the power of the state-endorsed map was reinforced by increasing commercially and publicly available aerial and satellite map imagery (ill. 17). At the same time, the widespread application of geographical information systems made people's lives steadily more efficient and hassle-free, even reading their maps for them. Despite ethical concern over the intrusive effects of GIS, and the false hand dealt by free, democratised yet state-controlled imagery, by the end of the century society largely believed itself comfortably on the map, able to acknowledge alternative maps but insulated from their subversive capabilities.

How did maps affect the twentieth-century experience? It is possible to look back upon the great themes of the century – war, peace, economics and social change – and see them as constructed by cartography. Total immersion in maps brought people under the influence of map-controlling government, military and commercial interests, but it also taught them in such a way that it became possible to see and to create alternatives. The power of the map proved too irresistible. There was hope that new mediums, particularly digital, might offer opportunities for alternative mappings. But in the twentieth century maps were democratic, based on the right of the state to control; they were ubiquitous, so long as people could afford them, and their degree of objectivity depended entirely upon the point of view.

This book was developed in conjunction with The British Library's 2016–17 exhibition of the same name. Although the book and exhibition are closely related and share many of the central ideas, the book is not a catalogue to the exhibition and can be read independently. The five chapters of the book mirror the five sections of the exhibition, for which the authors were asked to expand on particular aspects of them as they saw fit. Readers may notice that maps discussed in one chapter might have been equally well discussed in another. The ESSO-published road map discussed in the chapter on 'Maps and Money' could after all quite easily have sat in the chapter on 'Movement'. This illustrates how maps, like history, are never black and white (however hard they attempt to be), and are subject to different interpretations.

Finally, for reader, author and visitor, the vantage point of the twenty-first century is not some external objective position from which the mapping of the twentieth century can be observed free of bias. We remain products of the twentieth century; our interpretation is coloured by our experiences of the period because we are powerless to escape off our own map.

17
SPOT Controlled Image Base (CIB)
10-metre resolution satellite image
of Riyadh, Saudi Arabia, taken on
4 October 1986.
Centre National d'Etudes Spatiales
(CNES)

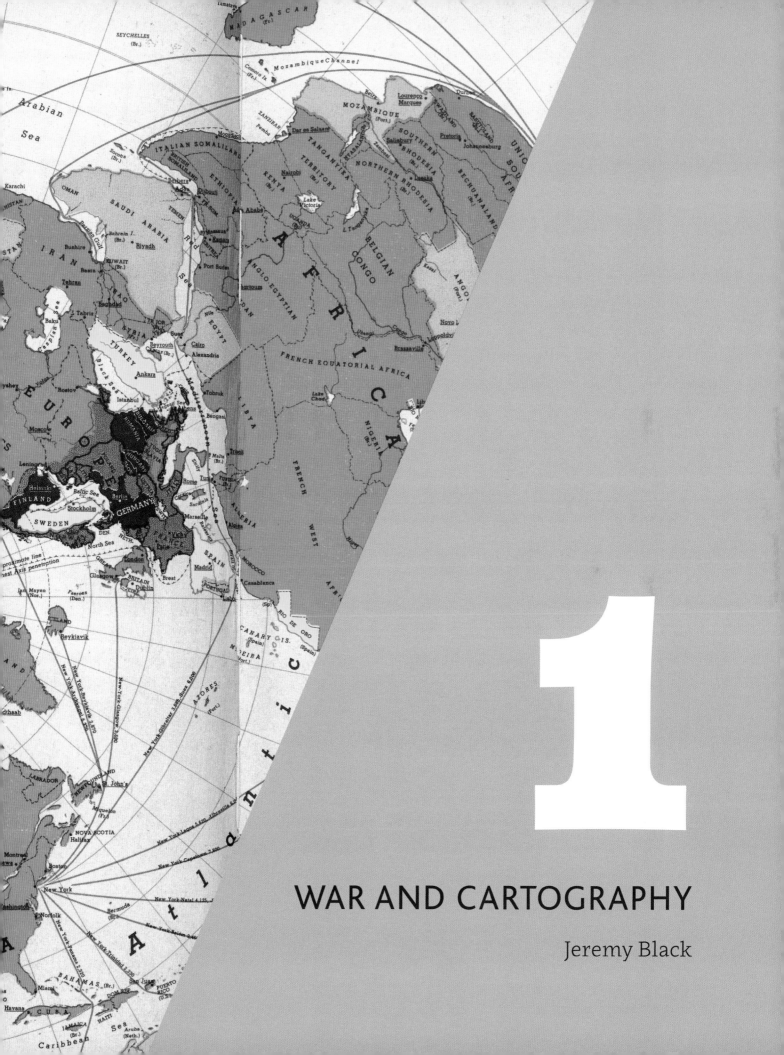

1

WAR AND CARTOGRAPHY

Jeremy Black

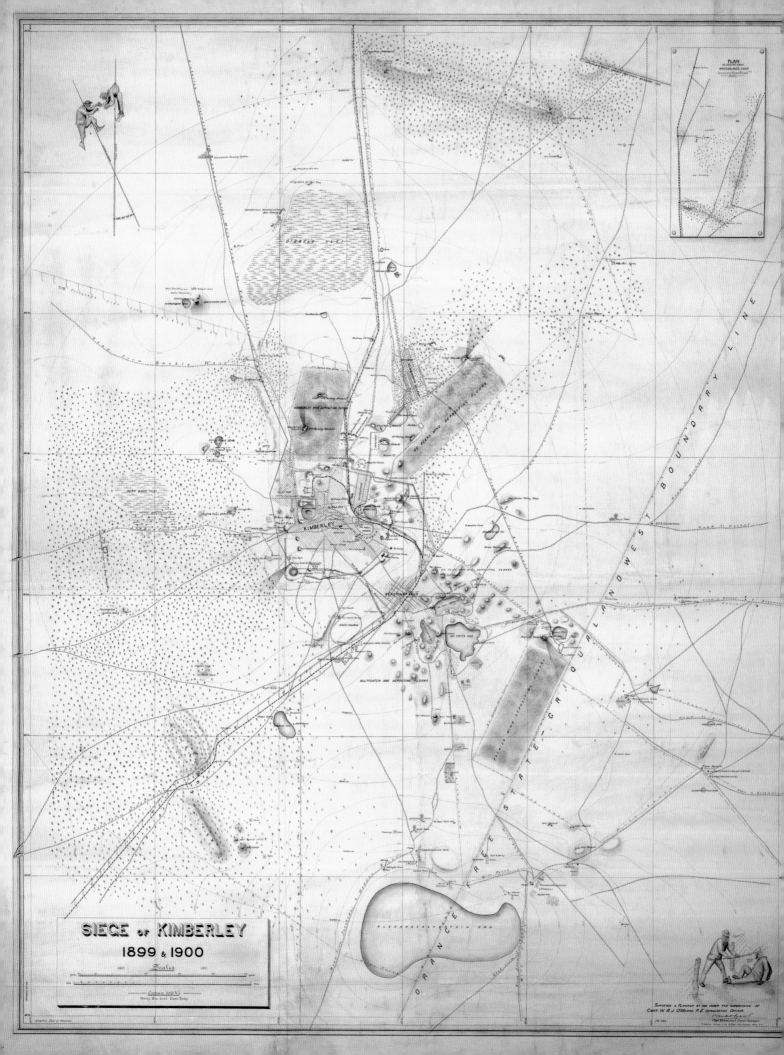

SIEGE of KIMBERLEY
1899 & 1900

The wars that took place in the twentieth century saw a great increase in the use of maps by the military, as well as public interest in them. Very large numbers of maps were produced for use by combatants and civilians alike, to an extent not seen previously. Maps have long been crucial in wartime in a range of ways, but most notably for operational purposes, news reporting and propaganda. Maps have also framed a particular perception of war among different audiences, and continue to do so. Finally, maps were important because developments in mapping technology during wartime have later been used in times of peace.

The First World War (1914–18) was unprecedented in terms of range, scale and intensity. As it unfolded there were therefore deficiencies with existing maps, and with the use of these maps for military operations. This proved challenging for the British and French attacking the German colonies in Africa; the continent was poorly mapped at the time, especially at a large or medium scale. This meant, for example, that in German East Africa (later Tanganyika, then most of Tanzania), the British navy knew in 1914 from signal intercepts that the German cruiser *Königsberg* was at Salale – but Salale was not marked on the naval charts, and it took eleven days to identify it. The lack of adequate maps of Africa also made it difficult to predict terrain or the location of watercourses.

Serious problems relating to maps also affected Turkish operations in the Caucasus in 1914, as well as German planning for a Turkish invasion of Egypt in 1915. More generally, the mapping of the Turkish empire in Asia – as opposed to that of Turkey's former empire in the Balkans – was poor. That posed particular problems when the Turkish campaign began in Mesopotamia (Iraq) and Palestine. The British had already faced serious deficiencies in mapping (and much else) at the time of the Boer War (1899–1902), and this had led to post-war reform in surveying (part of a more general reform process), which vastly improved British effectiveness. Subsequently the British devoted much attention to developing the cartography of the Middle East, using the survey systems already established in Egypt and India.

Alongside the problems posed by inadequate possession of maps of regions outside

18
Claude O. Lucas, *Siege of Kimberley 1899 & 1900*. Cape Town, 1900.
Add.MS 88906

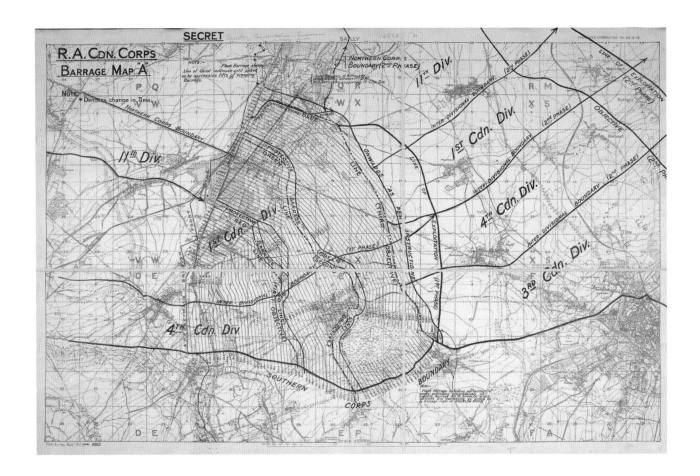

Europe, and the challenge of producing new ones, interpreting existing maps of Europe
was also difficult. This was seen, crucially, with the German invasion of Belgium and
France in 1914. Maps had always to be read in light of how swiftly the opposing forces
would be able to move, depending on the terrain, and how difficult they would find it to
mount an offensive. The French, for example, could readily use their rail system to move
troops from eastern France to support the defence of Paris, whereas the Germans faced
the difficulty of taking over rail systems in conquered areas, with different gauges of
track. Maps did not clearly communicate such issues, and did not enable commanders to
locate their units as accurately as necessary.

In the Battle of Ypres in the autumn of 1914, the particular needs of trench warfare
created demands that had not been anticipated by military cartographers. The French,
for example, had concentrated their military mapping on major fortified positions near
the border with Germany, such as Belfort (the site of a major German siege during the
Franco-Prussian War of 1870–1), only to discover that most of the fighting took place
nowhere near them, and that they were not prepared for the mobile warfare within
France in 1914. Nor were they prepared for the subsequent trench warfare, with the
associated need for detailed trench maps to enable the planning of effective defences
and successful assaults, which depended on infantry–artillery coordination (ills 19, 20).
Accurate surveying and mapping reduced the need for the registration of targets by guns

19 (above)
R.A. Cdn. Corps Barrage Map 'A'.
The map shows an area near
Bourlon, France. Field Survey
Battn., R.E., 1918.
Maps 14628.(20)

20 (right)
British trench map of an area south
east of Arras, annotated by Lt. John
Hodgson to show the aftermath
of a battle in 1918.
Maps CC.5.b.30

TRENCH MAP

REFERENCE
Scale 1:10000 — 176
TRENCH MAP
GERMAN TRENCHES shewn in RED
BRITISH " " Blue
Corrected to 17-8-17.
No 176

TRENCH in existence

under construction
or incomplete

impossible to dig
out

H.Q.

21
One of a set of relief models of
the Western Front 1:20,000 (also
known as Haig models), showing the
area round Ypres. Ordnance Survey,
Southampton, 1916–17.
Maps C.19.a-g

prior to attack, and made possible an element of surprise. This was crucial due to the strength of defensive firepower.

The scale of the increase in mapping during the First World War was impressive. When the British Expeditionary Force (BEF) was sent to France in 1914, with just one officer and one clerk responsible for mapping, the resulting maps were unreliable. However, by 1918 the survey organisation of the BEF had grown to about 5,000 men and produced more than 35 million map sheets. No fewer than 400,000 impressions were made in just ten days in August 1918. Ordnance Survey military topographic sections and field survey battalions surveyed and mapped different sectors of the Front from mid-1915, coordinated in France through the Ordnance Survey Overseas Branch from late 1917. The overwhelming majority of the maps produced by the Ordnance Survey during the war were large-scale battlefield sheets (ill. 21).

And this was not all: in May 1916, the General Staff urgently demanded from the

22
French trench map of the area around Berry au Bac. Service Topographique d l'Armée, Paris, 1918. Maps Y.72

Royal Geographical Society special thematic maps of the British sector of the Western Front, coloured to show relief and drainage features. These turned out to be vital in planning attacks.

The nature of trench warfare meant that maps were produced for the military at a far larger scale than those used for mobile campaigning (ill. 22). These large-scale maps required a high degree of accuracy to permit indirect fire, as opposed to artillery firing over open sights. This accuracy came in part from improvements in photogrammetry, and aerial information was extremely important. Cameras, mounted first on balloons, and later on aircraft, resulted in a breakthrough in surveillance; they were able to record fine details and to allow for scrutiny of the landscape from different heights and angles. Instruments for mechanically plotting maps from aerial photography were developed in 1908, while a flight over part of Italy by Wilbur Wright in 1909 appears to have been the first on which photographs were taken. The range, speed and manoeuvrability of aeroplanes gave them a great advantage over balloons.

With the First World War came air–artillery coordination. At first, maps did not play much of a role. A report in *The Times* on 27 December 1914 noted of the Western Front in France and Belgium:

The chief use of aeroplanes is to direct the fire of the artillery. Sometimes they 'circle and dive' just over the position of the place which they want shelled. The observers with the artillery then inform the battery commanders – and a few seconds later shells come hurtling on to, or jolly near to, the spot indicated. They also observe for the gunners and signal back to them to tell whether their shots are going to, whether over or short, or to right or left.

However, in time there were more static positions, with a need for heavier and more precise artillery fire in order to inflict damage on better-prepared trench systems. Both led to the use of maps as the key means of precision planning.

The invention of cameras capable of taking photographs with constant overlap was important in aerial reconnaissance and surveying, with the notable development of three-dimensional photographic interpretation. Maps worked to record positions as well as disseminate information. The ability to build up accurate models of opposing trench lines was but part of the equation. It was also necessary to locate the position of artillery in a precisely measured triangulation network, which permitted directionally accurate long-range artillery fire on particular coordinates. Lieutenant-Colonel Percy Worrall noted of the Western Front in April 1918: 'the artillery and machine-gun corps did excellent work in close co-operation … it was seldom longer than 2 minutes after I have "X-2 minutes intense" when one gunner responded with a crash on the right spot.' The volume of reconnaissance photography was such that the Germans were able to produce a new image of the entire Western Front every two weeks, and in turn rapidly produce maps that reflected changes on the ground (ill. 23).

More generally, there was an appropriation of the discipline of geography for the cause of war. In particular, the Royal Geographical Society (RGS) in London played a

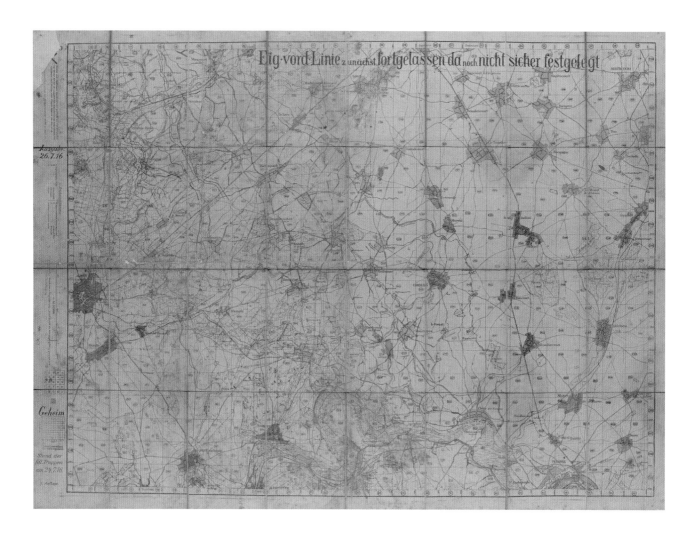

significant role as a cartographic agency that was closely linked to the British intelligence services. In 1914, it addressed the urgent task of producing an index of place names on the large-scale maps of Belgium and France that were issued to the British officers sent there. A four-sheet 1:500,000 wall map of Britain was also produced in order to help the War Office plan home defence strategies in the event of a German invasion. The RGS then pressed on to produce a map of Europe at the scale of 1:1,000,000. By the end of the war, over ninety sheets had been prepared by the RGS, covering most of Europe and the Middle East. The Ordnance Survey and many commercial businesses also played a major role.

The First World War stimulated public interest in maps, and newspaper readers expected them to accompany news stories, as during the Boer War. Maps were used to provide what were intended as objective accounts as well as propaganda, though the distinction between the two was not always clear. In the first case, many maps were used to locate areas of conflict. They provided a more valuable addition to the text than photographs, and were especially useful for the distant areas that were not covered by conventional atlases, and to show the level of detail necessary to follow trench warfare. Colour photography was not yet an established part of newspaper publishing, so black-

23

German trench map of the area north of Bray sur Somme, 1916.
National Army Museum, 1979-11-62

24

Journalistic map of the Western
Front, published in the *Nottingham
Evening Post* on 27 September 1915.
British Newspaper Archive

and-white maps were not overshadowed. However, making them genuinely helpful
was not easy. These simple maps reflected little of any tricky terrain or communication
difficulties (ill. 24).

At this time there was also a rapid production of atlases to satisfy consumer demand.
These included the *Atlas of the European Conflict* (Chicago, 1914), the *Daily Telegraph
Pocket Atlas of the War* (London, 1917), *Géographie de la Guerre* (Paris, 1917), *The Western
Front at a Glance* (London, 1917), *Petit Atlas de la Guerre et de la Paix* (Paris, 1918) and
Brentano's Record Atlas (New York, 1918). The value of geography during the war helped
to cement its development as a university subject. Honours schools were established
during or immediately after the war, including at Liverpool University in 1917, at the
London School of Economics and Aberystwyth University in 1918, and at University
College, London, and Cambridge and Leeds universities in 1919.

In the period between the world wars (1918–39), there were technological advances
that would affect mapping for war. In particular there were further developments in aerial
photography – including, in the 1930s, the development of colour and infrared film. Infrared
images show colours invisible to the naked eye, and can also help the viewer distinguish
between camouflaged metal, which does not reflect strongly, and the surrounding

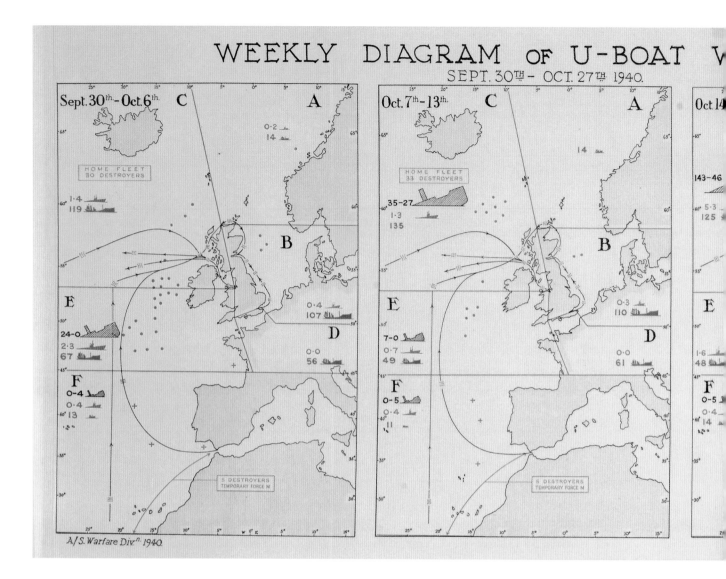

vegetation, which does. Meanwhile, despite the obvious advantages of colour film, black-and-white film still had a place as it showed stronger contrasts than colour.

In the Second World War (1939–45), the production and use of maps was on an even larger scale than in the First World War, which the numbers make clear: the British Ordnance Survey produced about 300 million maps for the Allied war effort, while the American Army Map Service produced more than 500 million. In addition, there were numerous other military mapping organisations producing maps for the Allies and more generally (ills 25, 26).

Once again, the range and scale of operations was such that there were shortages of the necessary maps. For example, a shortage of maps in military collections when the United States entered the war in 1941 necessitated extensive government borrowings, notably from the New York Public Library. In the United States, the need was advertised in popular magazines:

25
Weekly Diagram of U-Boat Warfare,
Sept. 30th–Oct. 27th, 1940.
Printed by the Anti-Submarine
Warfare Division.
The National Archives MFQ 1/583 (26)

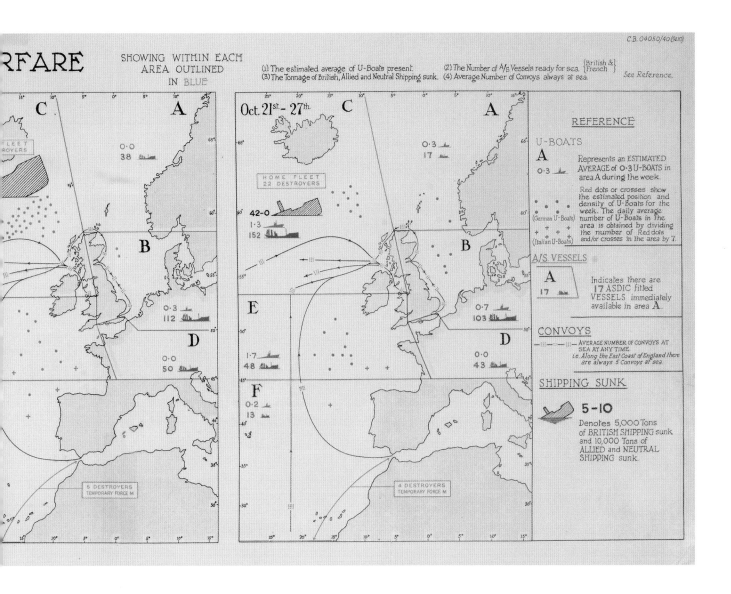

The War Department, Army Map Service, is seeking maps, city plans, port plans, place name lexicons, gazetteers, guide books, geographic journals, and geologic bulletins covering all foreign areas outside the continental limits of the United States and Canada. Of particular interest are maps and guide books purchased within the last ten years, including maps issued by the U.S. government and the National Geographic Society.

The situation was swiftly remedied. Units in the field were provided with plentiful maps, and were also able to produce them in response to new opportunities and problems; cartographic expertise was at a premium. Walter Ristow from the Map Division of the New York Public Library became head of the Geography and Map Section of the New York Office of Military Intelligence. Armin Lobeck, Professor of Geology at Columbia University, who had produced a major study of geomorphology in 1939, produced maps and diagrams in preparation for Operation Torch, the successful American amphibious invasion of French North Africa in 1942. Lobeck also produced a set of strategic maps for Europe. Specific

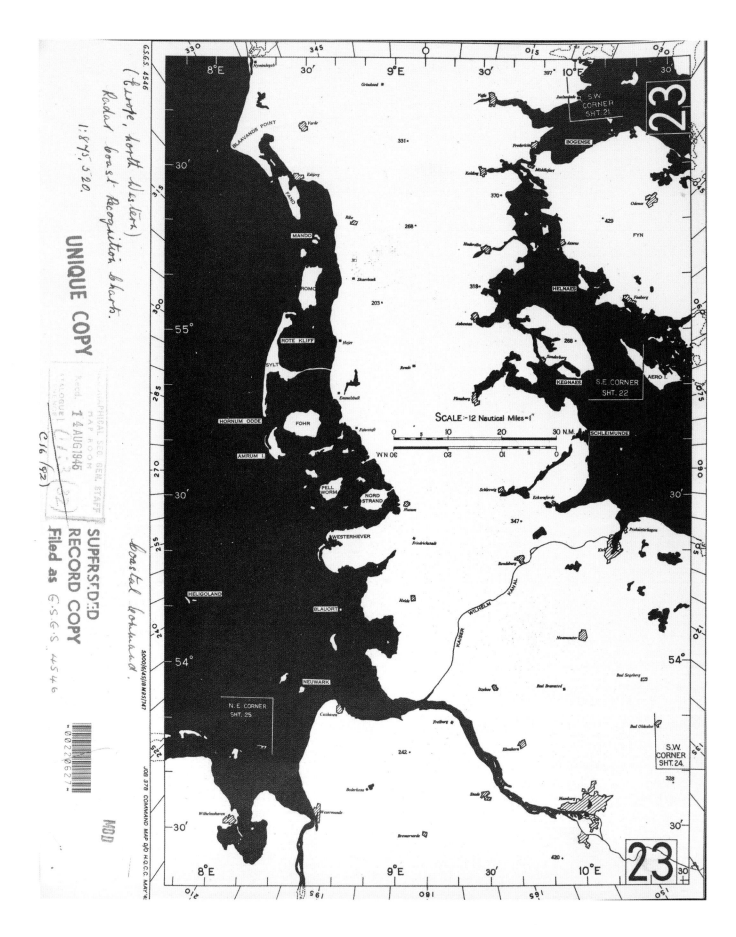

maps and atlases were also requested, such as the *Atlas of Svalbard* (1927) in June 1942: the island of Spitsbergen was important due to its coal, gypsum and asbestos mines and as a highly useful weather station for both sides. The data produced was itself mapped. Meteorological analysis and mapping became particularly important during the war, notably for air and sea operations. For example, by 1944 the Allies were employing radar and radio direction finders to track the hydrogen balloons that were used to indicate wind speed and direction.

The need to coordinate air and land, air and sea, and land and sea operations resulted in the increased complexity of many maps, to help with planning in three dimensions. This was especially true for bombing and tactical ground support from the air, and for airborne attacks by parachutists and gliders. The development of the operational dimension of offensives, especially in the Soviet army in 1943–5 (notably Operation Bagration in 1944), hinged on an ability to seize and retain the initiative and to outmanoeuvre opponents. This in turn depended on staff being well informed about the distances and locations involved. Long-range bombing, such as that of Japan by the Americans in 1942 and 1944, necessitated particularly accurate maps as it pushed the limit of the bombers' flying range. The American Office of Military Intelligence acquired and reproduced the *Atlas of Japan* (1931), produced by the Japanese Land Survey Department, as well as many other maps of Japan, and also created a visual index to the atlas plates using a 1937 railway map of Japan with a hand-drawn grid showing the locations of the individual plates.

To help bombers, existing printed maps were acquired by all powers, supported by the products of various surveillance activities. For example, to guide their bombers to targets in Britain, the Germans used British Ordnance Survey maps enhanced with photographic reconnaissance information (ill. 27). Aeronautical charts for aircraft became increasingly sophisticated, tailored to different tasks and speeds. In the United States, five basic charts, each with a distinctive scale, were developed during the war: for flight planning, cruising, descent, approach, and landing and taxiing. At the same time radar and radio navigation systems were adopted, with radio navigation charts employed to represent transmission stations.

Aerial photography improved significantly during the Second World War, thanks to better cameras that offered improved magnification. Equally significantly, photo-interpreters improved their analysis capabilities of the resulting film. The German attack on the Soviet Union in 1941 was preceded by long-range reconnaissance missions by aircraft flying at high altitude. Meanwhile, facing the need to operate in the poorly mapped Pacific, the Americans made extensive use of photo-reconnaissance, not least for mapping invasion beaches.

Secrecy has always been at the heart of mapping for war. Aside from producing their own maps, in the Second World War governments also restricted the distribution of maps that might help enemies: in the United States, this meant the topographic maps produced by the Geological Survey and the nautical charts produced by the Coast and Geodetic Survey. Conversely, espionage was often focused on gaining access to secret plans and maps. In the run-up to D-Day, a British MI6 agent stole the plans of the Atlantic Wall, the German defences, while other British agents provided reliable maps of the landing beaches from reconnaissance.

26

Radar coast recognition chart of part of the North German coast. Produced by the Air Ministry, Coastal Command, 1945. Maps MOD GSGS 4546 (26)

Bildskizze
Liverpool-Birkenhead

Maßstab etwa 1 : 40 000

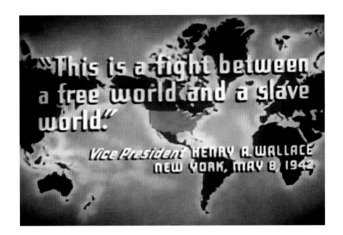

27 (left)
Aerial photograph of Liverpool with targets marked in red, from a booklet of British industrial targets for Luftwaffe operations. Luftwaffe, Berlin, 1940.
Maps 46.d.1

28 (above)
Two screenshots from the *Why we Fight* film series (1942–5), directed by Frank Capra

29 (next page)
British propaganda leaflet, in Arabic, showing 1942 Allied advances in Libya and Egypt, with a portrait of Commander-in-Chief Harold Alexander.
Maps CC.5.b.43

As in the First World War, maps were used both to convey news and for propaganda. Once again, newspapers printed large numbers of maps. However, the rather two-dimensional notion of an easily rendered front line was not always appropriate; this was especially true of the Pacific, where a number of important Japanese bases such as Rabaul and Truk were simply bypassed by the Americans in 1943–5 thanks to superior air and sea power. Such a fluid and fast-moving situation could not be readily captured on a map. Similarly, the detrimental effects of autumn rains, winter ice and spring thaws on the roads at the Eastern Front could not be represented.

Maps had a major role to play in propaganda. *Why We Fight* – a series of seven motivational films made from 1942 to 1945, commissioned by the US government and directed by Frank Capra (ill. 28) – included *Prelude to War* (1943). This depicted a hemisphere of light and another of dark dictatorship, while the maps of Germany, Italy and Japan were transformed into menacing symbols. Other prominent propaganda maps included those produced for a French audience by the German propaganda department based in Paris, one of which depicted Churchill as an octopus reaching out to attack the French empire, with the attacks being bloodily repelled. Morale-shredding maps were published and distributed, sometimes dropped from airplanes, for the attention of civilians or retreating enemy troops (ill. 29).

As in the First World War, commercial atlases were produced, including *Atlas of the War* (Oxford, 1939), *The War in Maps* (London, 1940) and *The War in Maps: An Atlas of The New York Times Maps* (New York, 1943). The aerial dimension also encouraged the use of particular perspectives and projections for maps. The innovative cartographer Richard Edes Harrison had introduced the perspective map to American journalism in 1935. His orthographic projections and aerial perspectives brought together the United States and distant regions, and were part of a worldwide extension of American geopolitical concern and military intervention. Indeed the role of aircraft, dramatically demonstrated to American civilians by the Japanese surprise attack on Pearl Harbor in December 1941, led to a new sense of space – which reflected both vulnerability and the awareness of new geopolitical relationships. Thus, *The Esso War Map* (1942) emphasised the wartime value of petroleum products – 'Transportation – Key to Victory' was the theme of the text – and also provided an illuminated section, 'Flattening the Globe', showing how the globe

الجنرال الكسندر القائد العام فى الشرق الأوسط
الذى نظم خطة زحف الجيش الثامن
وانتصاره فى ليبيا

الجيش الثامن يتقدم نحو طرابلس وعلى وجوه جنوده امارات الثقة بالنصر والاغتباط بفوزم الساحق .

الدخان يتصاعد من إحدى السفن التجارية قرب ميناء بنغازى التى ضربها الحلفاء ضرباً متواصلا بالقنابل وقد احتلوها بعد ١٧ يوما فقط من بدء زحفهم

النّصر

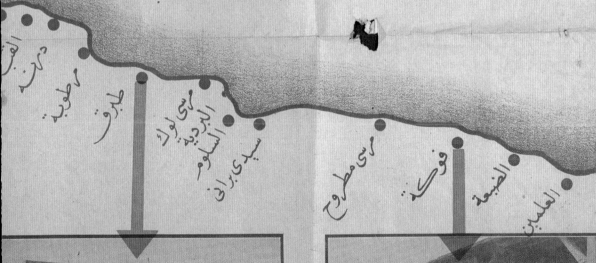

طبرق · مطروح · درنة · من · اقطوبية · اليرديه · ليوك · التلوم · سيدي براك · لاي مطروح · فوكة · الضبعة · العليين

دبابات الجيش الثامن تدخل طبرق بعد انقضاء تسعة أيام فقط على تحطيم خطوط دفاع الألمان

جنديان الماليان يستظلان لحدى بريطاني في فوكا . وقد سقطت فوكة في الأيام الأولى من الزحف البريطاني الباهر

NEWSMAP
Prepared and distributed by ARMY ORIENTATION COURSE.
Special Service Division Army Service Forces,
WAR DEPT. 26581 Pentagon Bldg., Washington, D. C.
★ U. S. GOVERNMENT PRINTING OFFICE: 1943 — O26110

TARGET Berlin

This map is a photographic view of the world with the center at Berlin. Thus, with the detachable scale, distances can be measured along any line running thru Berlin. It should be noted that an inch at the center represents less mileage than an inch closer to the edges. The detachable scale has been designed to compensate for this and should be used only with the center on Berlin.

MAP

The photographic process used in making this map makes all distances measured with the tape approximate only. Distances are shown in statute miles. Lines between key cities do not represent regular air routes in all cases. They show distances between points that do not fall on a line going thru the center of the projection.

Cut Along Dotted Line

Cut Along Dotted Line

Scale Ⓐ

Scale Ⓑ

This scale is correct only when the center is placed at Berlin.

becomes a map. North America was positioned centrally, which made Germany and Japan appear as menaces from the east and west. The map included sea and air distances between strategic points such as San Francisco and Honolulu. Using the same technique, F. E. Manning's globular maps of 1943 positioned Berlin and Tokyo as vulnerable targets at the centre of the globe (ill. 30). Public engagement was a clear theme.

Popularising geopolitics in a radio speech, President Franklin Delano Roosevelt's address to the nation on 23 February 1942 referenced a map of the world to explain American strategy. He had earlier suggested that potential listeners obtain such a map, which led to massive demand and also to increased publication in newspapers. Already, on 9 September 1939, Rand McNally had announced that more maps had been sold at its New York store in the first twenty-four hours of the war than during all the years since 1918. There was another upsurge in sales after Pearl Harbor, when America entered the war. Roosevelt, who obtained his maps from the National Geographic Society, created a map room in the White House. For Christmas 1942, he was given a huge fifty-inch globe, manufactured under the supervision of the Map Division of the Office of Strategic Services and the War Department and presented by the Chief of Staff, General Marshall.

The task of explaining engagement with distant regions posed a problem, but also offered opportunities for innovation in both conception and presentation. The film *Man for Destruction* (1943) depicted the German politician Karl Haushofer, purportedly close to the Nazi regime, explaining global geopolitics in front of a map centred on the North Pole – an exposition of a threat implicit in linking different parts of the world, and a suggestion of what the American response should be. Richard Edes Harrison also

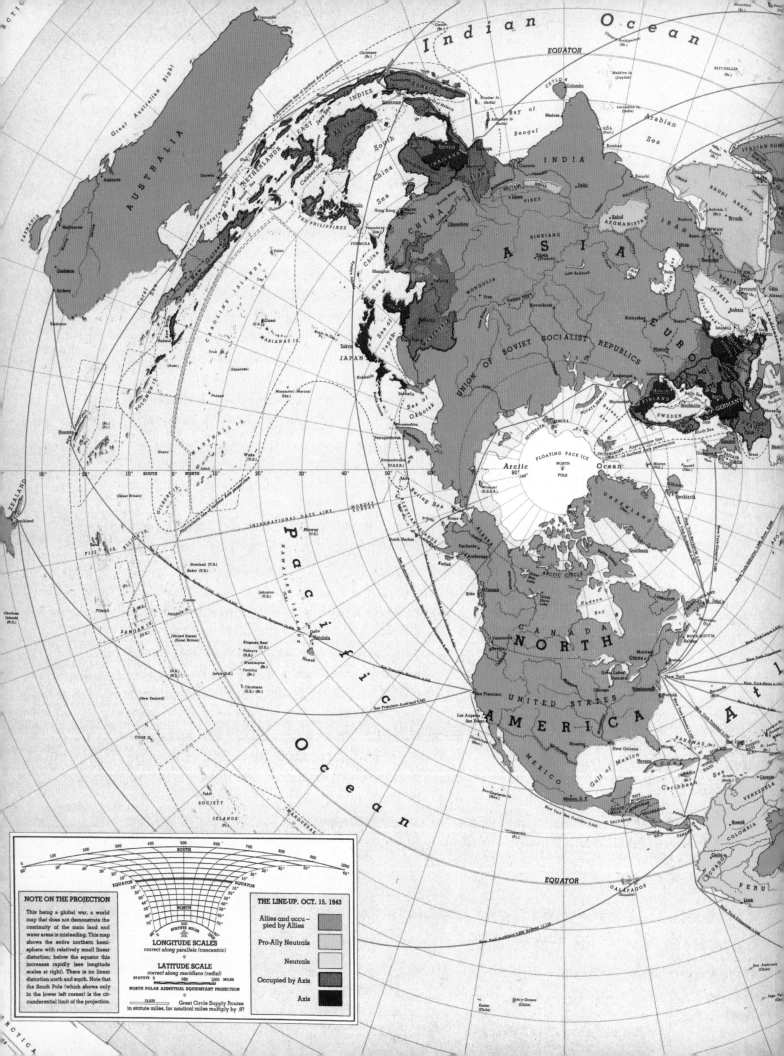

THE LINE-UP, OCT. 15, 1943

Allies and occupied by Allies

Pro-Ally Neutrals

Neutrals

Occupied by Axis

Axis

NOTE ON THE PROJECTION

This being a global war, a world map that does not demonstrate the continuity of the main land and water areas is misleading. This map shows the entire northern hemisphere with relatively small linear distortion; below the equator this increases rapidly (see longitude scales at right). There is no linear distortion north and south. Note that the South Pole (which shows only in the lower left corner) is the circumferential limit of the projection.

LONGITUDE SCALES
correct along parallels (concentric)

LATITUDE SCALE
correct along meridians (radial)

STATUTE 0   500   1,000   MILES

NORTH POLAR AZMUTHAL EQUIDISTANT PROJECTION

Great Circle Supply Routes
in statute miles, for nautical miles multiply by .87

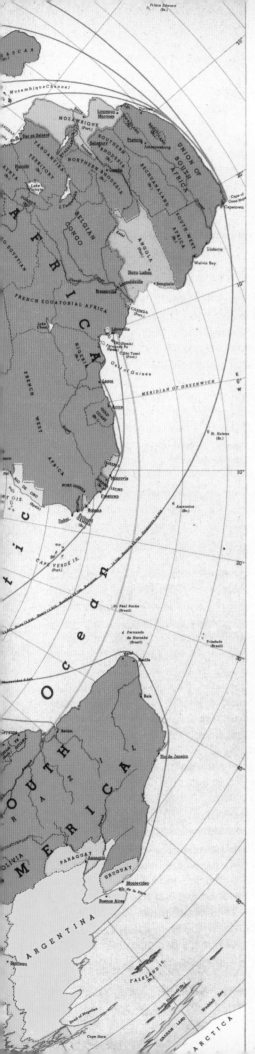

One World, One War

THE north polar sea is essentially the center of our world. To one side is North America, to the other Asia and its offshoot, Europe—triangular continental masses pointing south toward distant seas and toward barren Antarctica. But in the north, the bases of the two triangles almost touch. Bering Strait, the point of separation in the Pacific, is only fifty-odd miles wide; and it is difficult to judge whether Iceland, the steppingstone in the Atlantic, belongs truly to the new or the old hemisphere.

Indeed, without a polar sea center, there might be one globe; but there would hardly be one world. If the continents were equidistantly separated, it would be very possible to have six wars and very difficult to make a single one out of them. In such case, almost all areas of the globe would have equal strategic value, and there would be no single trade routes of outstanding importance. It is the nearness of the northern continents that makes certain areas vital to world trade and to world security. Furthermore, there is a very close relationship between the geographical proximity of the various land masses and their population density. Over 90 per cent of the world's people live in lands north of the equator, essentially because it has always been shorter to travel close to the polar sea than to travel around the southern oceans.

The map to the left is in part a war map. It shows the World War II line-up of nations, and it traces the battle fronts and supply lines of the various arenas. It is a map of the problems and the opportunities of fighting all over the world all at once. While it includes obvious distortions, which increase toward the south, it serves as an excellent all-over strategy map. It is a continuous map that shows the world in one unbroken piece. Furthermore, it is centered within the great triangle formed by the world's power centers. This triangle is shown with a minimum of distortion. The map, therefore, is an index of the atlas that it introduces.

The maps that follow describe in greater detail various facets of the Northern Hemisphere world. They do not attempt more than a passing view of Africa, South America, and Australia; they do not aim to catalogue the place names of such spots as Cyprus. They are practical lesson maps; they do not seek to be encyclopedic or to present the globe as a subject of abstract study. Rather they emphasize the too-long-forgotten realities of world geography.

COMPLETE AZIMUTHAL EQUIDISTANT PROJECTIONS

 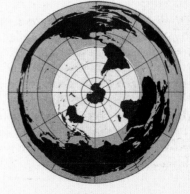

Centered on North Pole   *Centered on South Pole*

33
Voici les Bases Americaines dans le Monde: Que est l'agresseur? Qui Menace? (Here are the American Bases throughout the World: Who is the Aggressor? Who is the Threat?)
Parti Communiste Francais, 1951

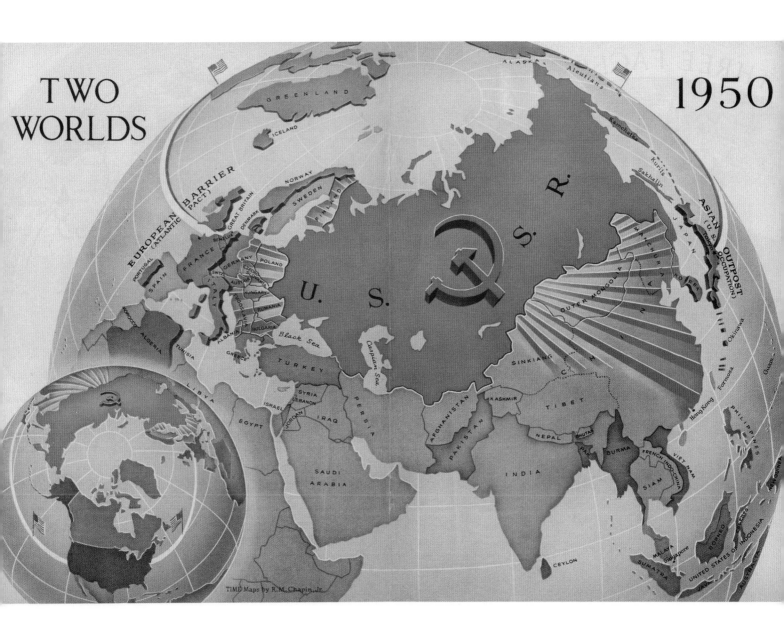

TWO
WORLDS

1950

34
Robert M. Chapin, *Two Worlds 1950*.
Published in *Time Magazine*,
2 January 1950

produced a map centred on the North Pole, a polar azimuthal projection, with the USA shown in a key position (ill. 32). The preface to Harrison's *Look at the World: The Fortune Atlas for World Strategy* (New York, 1944), an atlas that reproduced his maps from the magazine *Fortune*, explained that it was intended 'to show *why* Americans are fighting in strange places and *why* trade follows its various routes. They [the maps] emphasise the geographical basis of world strategy.' Harrison's maps put the physical environment before national boundaries, and also reintroduced a spherical dimension, offering over-the-horizon views: an aerial perspective that does not exist in nature but that captured physical relationships, as in his 'Europe from the Southwest', 'Russia from the South', 'Japan from Alaska' and 'Japan from the Solomons'. The first edition of the atlas sold out rapidly, and Harrison's techniques were widely copied and his maps used in training pilots.

The reporting and presentation of war, notably the dynamic appearance of many war maps – for example those in *Fortune*, *Life* and *Time*, with their arrows and general sense of movement – helped to make geopolitics present and urgent. Far from the war appearing to American readers as a static entity, and at a distance, it was seen as in flux. The maps also made the war seem to encompass the spectator visually, through images of movement, and in practice, by spreading in their direction. The orthographic projection used for the map entitled *The Aleutians: Vital in North Pacific Strategy*, published in *The New York Times* on 16 May 1943, depicted the American island chain that had been attacked by the Japanese as the centre in a span stretching from China to San Francisco. This presentation made apparent the islands' potential strategic importance. The Office of War Information followed up with *A War Atlas for Americans* (1944), which offered perspective maps.

The Cold War between Eastern and Western power blocs descended in 1948. It was presented in geopolitical terms, both for analysis and for rhetoric. As during the Second World War, a sense of geopolitical challenge was used to encourage support for a posture of readiness, indeed of immediate readiness. The sense of threat was expressed in map form, with both the United States and the Soviet Union depicting themselves surrounded and threatened by the alliance systems, military plans and subversive activities of their opponents. These themes could be clearly seen not only in government publications, but in those of other organisations too (ills 33, 34). The dominant role of the state helps to explain this close alignment in the case of the Soviet Union and its Communist allies. In the United States, there was also a close correspondence between governmental views and those propagated in the private sector, not least in the print media.

A sense of threat was apparent in the standard map projection used in the United States. The van der Grinten projection, invented in 1898, continued the Mercator projection's practice of exaggerating the size of the temperate latitudes: Greenland, Alaska, Canada and the Soviet Union all appeared larger than they were in reality. This projection was used by the National Geographic Society of America from 1922 to 1988; its maps were the staple of American educational institutions, the basis of maps used by newspapers and television and the acme of public cartography. In this projection, a large Soviet Union appeared as a menace to the whole of Eurasia – a dominant presence that required containment.

35

US president John F. Kennedy uses
a map of Laos as a prop to explain
the military situation to a press
conference on 25 March 1961

However, rather than use these examples simply to decry American views of the
time, it is necessary to point out that Soviet expansionism was indeed a serious threat,
certainly to the states of Central and Eastern Europe – and the geopolitical challenge
was particularly acute due to the Soviet Union being both a European and an Asian
power. The Soviets meanwhile employed their own cartographic imagery and language,
casting the United States as hostile.

Carrying forward the tradition of President Roosevelt's use of maps to support
his fireside chats over the radio, President John F. Kennedy, in a press conference on
23 March 1961, employed maps when he focused on the situation in Laos (ill. 35), a
French colony until 1954, where the Soviet and North Vietnamese-backed Pathet Lao
were advancing against the forces of the conservative government:

*These three maps show the area of effective Communist domination as it was last
August, with the colored portions up on the right-hand corner being the areas held*

and dominated by the Communists at that time. And now next, in December of 1960, three months ago, the red area having expanded – and now from December 20 to the present date, near the end of March, the Communists control a much wider section of the country.

The use of red dramatised the threat, as did the depiction on the map of the countries bordering Laos: Thailand, Cambodia, South Vietnam and Burma. In this way the domino theory, or the theory of Communist advance in stages, was used to justify the deployment of 10,000 American Marines based in Okinawa from 1945.

Although traditional methods of information gathering continued (as in the compilation of Soviet maps of British and other towns from various published sources, ill. 36), mapping for war was affected by the changing character of surveillance and information. This was influenced by the greater speed and range of aircraft. In particular, in the 1950s the Americans deployed first a version of the B-47 and then the U-2, which could take accurate photographs from a greater height and which seemed safe from interception.

The extent to which the United States and Britain relied on air power for strategic ends also played a part: both as the delivery system for atomic weaponry and in order to offset the marked numerical advantage of the Red Army in Europe. The speed of jet aircraft and their reliance on a far smaller crew than long-range bombers – often just a single pilot – necessitated the development of cockpit-based moving-map displays, mounted on the instrument panel. These went through a process of development and were particularly suited to computer operation.

Aircraft as a source of information and power projection were supplemented by satellites, projected into orbit by rockets; the first satellite, the Soviet *Sputnik*, was launched in 1957. Orbiting satellites offered the potential for obtaining material, for the radio dispatch of images and for the creation of a global telecommunications system. In 1960 the first pictures were received from *Discoverer 13*, one of the earliest American military photo-reconnaissance satellites. The overlap between photography and mapping became increasingly pronounced; regular satellite images provided the opportunity to map change on the ground, including the construction of missile sites. Satellites were too high to be shot down by a ground-based missile, as the U-2 American surveillance aircraft had been in 1960. Moreover, they offered the possibility of frequent overflights, so providing more data.

This satellite information also came to serve as the basis for enhanced weaponry. The US Department of Defense developed a Global Positioning System (GPS) that relied on satellites, the first of which was launched in 1978. Meanwhile automatic aiming and firing techniques depended on accurate surveying. 'Smart' weaponry, such as guided bombs and missiles that are used to this day, makes use of precise mapping to follow predetermined courses to targets actualised for the weapon as a grid reference. Computerisation was to become highly important in the deployment of information – cruise missiles, for example, use digital terrain models of the intended flight path. Not surprisingly the Soviet Union sought to match the American GPS system with its own

36 (left)
Soviet military map of Brighton and Hove on the south coast of England, one of a number of Soviet plans of foreign towns unearthed in abandoned military map depots in Latvia and other former Soviet states in 1992. Braiťon i Khov. Generalńyĭ Shtab, 1990. Maps X.4609

37 (next page)
Escape and evasion map of Iraq, printed on Paxar, for the use of British pilots. Ministry of Defence, 1990. Maps X.4578

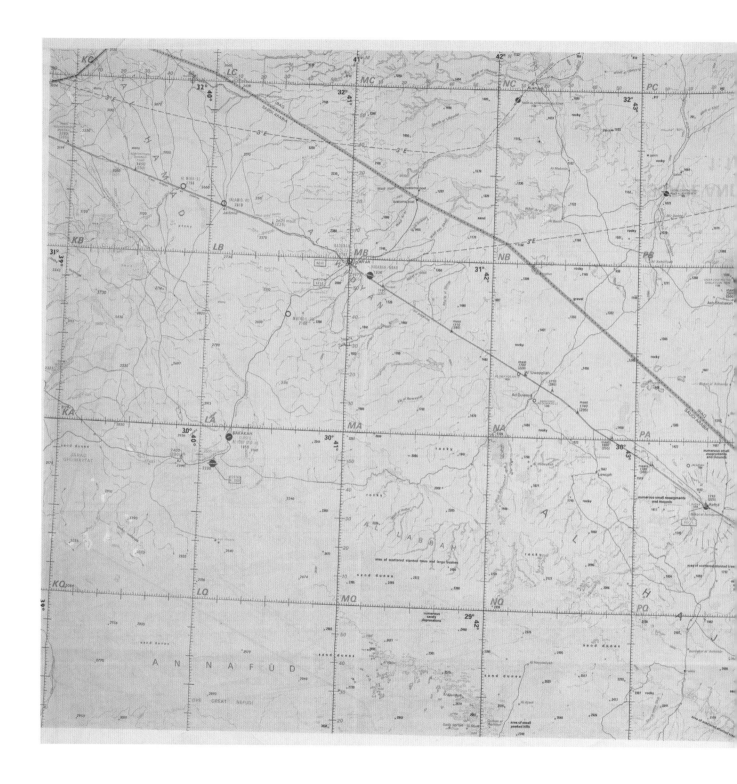

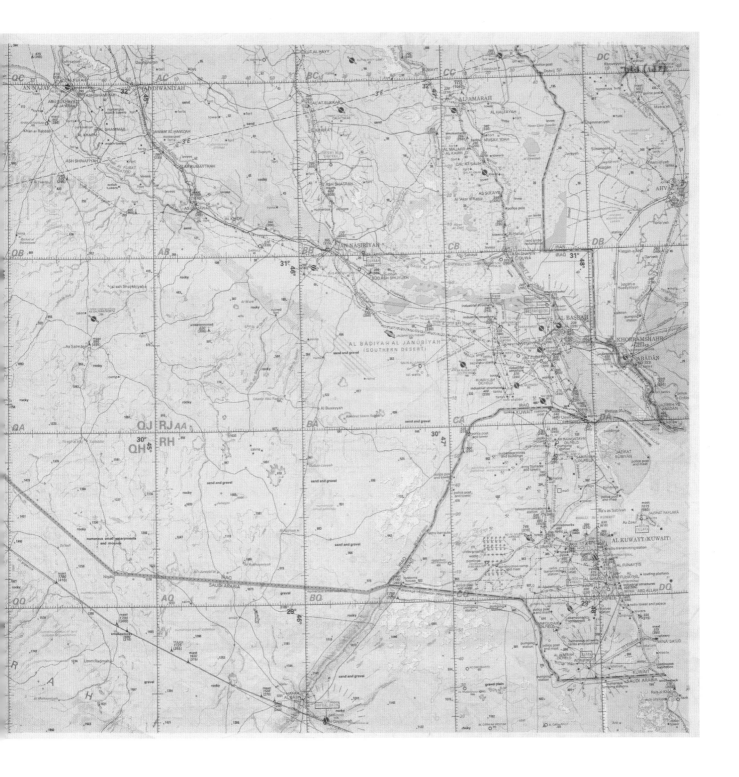

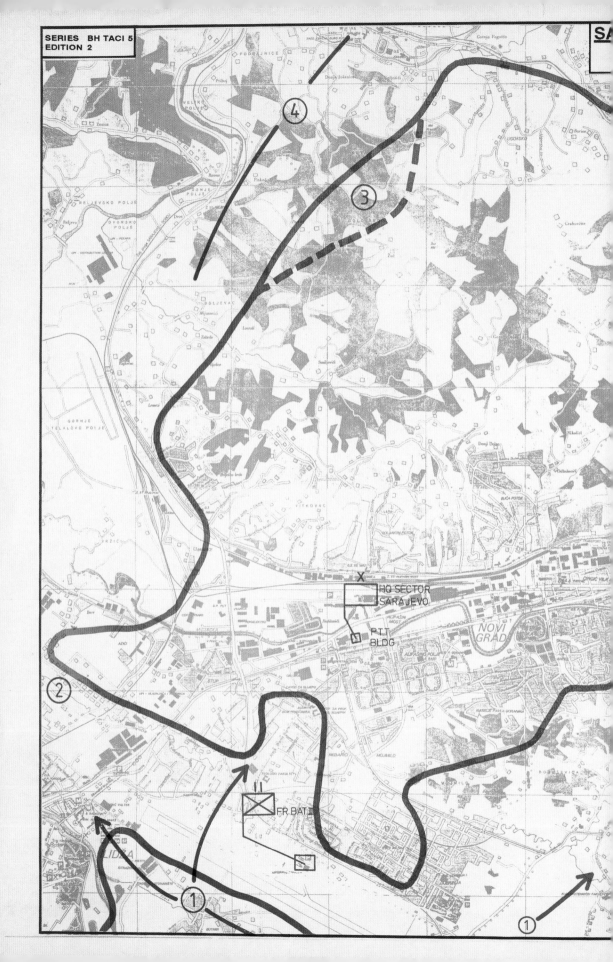

SERIES BH TACI 5
EDITION 2

SA

④

③

②

①

X HG SECTOR
SARAJEVO

PTT
BLDG

II
⊠ FR BAT II

①

①

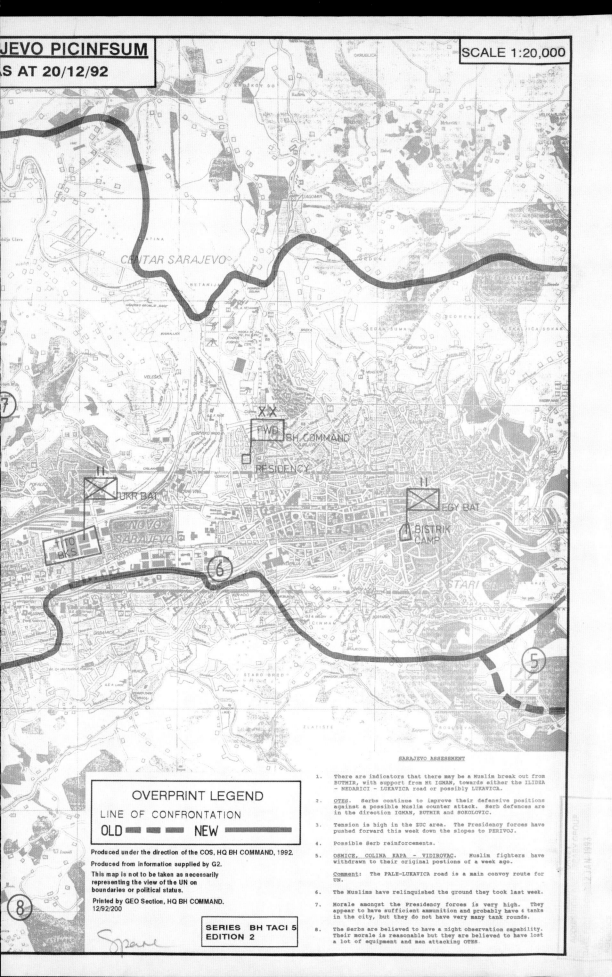

SCALE 1:20,000

OVERPRINT LEGEND

LINE OF CONFRONTATION

OLD ▬ ▬ ▬ ▬   NEW ▬▬▬▬

Produced under the direction of the COS, HQ BH COMMAND, 1992.

Produced from information supplied by G2.

This map is not to be taken as necessarily
representing the view of the UN on
boundaries or political status.

Printed by GEO Section, HQ BH COMMAND.
12/92/200

SERIES BH TACI 5
EDITION 2

SARAJEVO ASSESSMENT

1.   There are indicators that there may be a Muslim break out from
     BUTMIR, with support from Mt IGMAN, towards either the ILIDZA
     - NEDARICI - LUKAVICA road or possibly LUKAVICA.

2.   OTES. Serbs continue to improve their defensive positions
     against a possible Muslim counter attack. Serb defences are
     in the direction IGMAN, BUTMIR and SOKOLOVIC.

3.   Tension is high in the ZUC area. The Presidency forces have
     pushed forward this week down the slopes to PERIVOJ.

4.   Possible Serb reinforcements.

5.   OSMICE, COLINA KAPA - VIDIROVAC. Muslim fighters have
     withdrawn to their original postions of a week ago.

     Comment: The PALE-LUKAVICA road is a main convoy route for
     UN.

6.   The Muslims have relinquished the ground they took last week.

7.   Morale amongst the Presidency forces is very high. They
     appear to have sufficient ammunition and probably have 6 tanks
     in the city, but they do not have very many tank rounds.

8.   The Serbs are believed to have a night observation capability.
     Their morale is reasonable but they are believed to have lost
     a lot of equipment and men attacking OTES.

Global Navigation Satellite System, which in 1996 reached its full design specification of twenty-four satellites.

In the 1991 Gulf War with Iraq (the First Gulf War, ill. 37), the Americans used cruise missiles. Moreover, precise positioning devices interacted with American satellites in a GPS that was employed with success by Allied tanks. In addition, satellite information helped in the rapid production of photo-maps, while geographical information system (GIS) software provided instantaneous two- and three-dimensional views of battlefields.

Alongside this high-specification mapping continued to exist the reality of more ordinary maps used for struggles at the local level, particularly by states that lacked this cutting-edge capability. The limited mapping of much of the world for much of the century, and notably its first two-thirds, could have been a factor. In 1963–6, for instance, the maps used by the British military in their 'confrontation' (low-level conflict) with Indonesia in Borneo were largely blank for the Indonesian side of the frontier with the former British territories, now part of Malaysia, that the British were protecting. The British had already extensively mapped areas of military activity in the defence of their empire, notably Malaya and Kenya. Ground surveying continued to be supplemented by aerial photography.

The situation was far more challenging for insurrectionary groups, as they did not have access to the modern facilities used in mapping. However, their need for maps was more limited because they tended to have more local knowledge, which brought them a huge tactical advantage. The only solution for any occupying force was to have as full, accurate and suitable a map coverage as possible (ill. 38).

The mapping of insurgency and counterinsurgency struggles, and indeed terrorism and counterterrorism, was problematic, and this is an issue that continues to this day. In this type of warfare, the notion of control over territory is compromised by the involvement of forces that cannot be readily described in terms of conventional military units. They seek to operate from within the civilian population for cover and sustenance, and also in order to deny their opponents any unchallenged control over populated areas.It is extremely difficult to map a situation of shared presence – one in which military or police patrols move unhindered, or suffer occasional sniping and ambushes and have to consider mines, but otherwise control little beyond the ground they stand on. Aerial supply and attack capabilities further complicate the situation. Thus, front lines dissolve.

Many conflicts today occur as forms of civil war and therefore within countries – for example Afghanistan, Iraq and Syria – so the situation is complex and not one readily addressed in terms of cutting-edge capability, whether in relation to weaponry or information availability and cartographic methods. There is no sign that this situation is changing.

At the same time, it may be the case that war develops in a different direction, with more conventional conflicts between great powers being revived. In the mid-2010s, there were major points of tension between China and Japan and between Russia and NATO. These very much suggested that 'wars among the people' might not be the sole theme or

38 (previous page)
Sarajevo Picinfsum as at 20/12/92. Update map of the positions and actions relating to the siege of Sarajevo. Bosnia-Herzegovina (BH) Command, Sarajevo, 1992

even – perhaps – the major one. This possibility poses significant problems for mapping, not least as technology develops. For example, the introduction of enhanced Chinese anti-ship missiles, notably the DF-26, a ballistic missile with a range of 3,000–4,000 kilometres, raises issues, not least in terms of how best to depict vulnerability. This dynamic potential means that mapping war will remain exposed to repeated changes

The tensions between military and public, functional and expository, and secret and open in the production, presentation and understanding of maps will continue to be significant – but the contexts, both military and public, will change rapidly.

MAP OF THE WORLD
PETERS PROJECTION

A map which represents countries accurately
according to their surface areas.

SCALE 1: 1,230,000,000 MILLION

One square centimetre on the map = 123,000 square kilometres
in nature.

THIS PROJECTION SHOWS COUNTRIES IN PROPORTION TO
THEIR RELATIVE SIZES. IT IS BASED UPON ARNO PETERS'
DECIMAL GRID WHICH DIVIDES THE SURFACE OF THE EARTH
INTO 100 LONGITUDINAL FIELDS OF EQUAL WIDTH AND 100
LATITUDINAL FIELDS OF EQUAL HEIGHT. IT TREATS THE
RECTANGLES AROUND THE EQUATOR AS SQUARES AND
BUILDS THE OTHER RECTANGLES ONTO THESE IN
PROPORTION TO THE AREAS THEY REPRESENT. THE ZERO
MERIDIAN ON THIS SYSTEM IS COMBINED WITH A PROPOSED
NEW INTERNATIONAL DATE LINE.
THIS NEW DECIMAL GRID IS ONLY INDICATED, HOWEVER, ON
THE OUTER BORDER OF THE MAP. THE GRID MARKED ON THE
MAP ITSELF IS BASED ON THE TRADITIONAL 180 DEGREE
DIVISION AND THE PRESENT DATELINE IS INDICATED WITH A
DOTTED LINE.

COPYRIGHT BY AKADEMISCHE VERLAGANSTALT, FL-9490 VADUZ, ÄEULESTR. 56.
ENGLISH VERSION BY OXFORD CARTOGRAPHERS, OXFORD, UK.

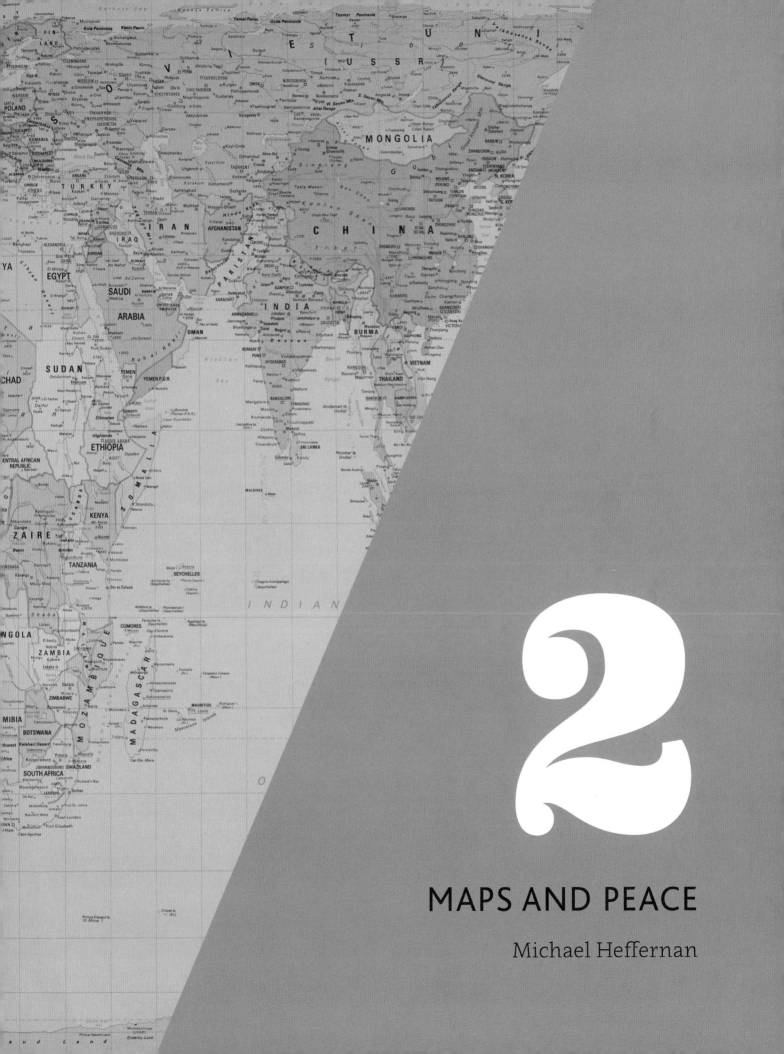

2

MAPS AND PEACE

Michael Heffernan

The role of maps in the visual culture of the twentieth century has been widely debated in recent years.[1] As Jeremy Black shows in Chapter 1, attention has focused to a large extent on the geopolitical, geostrategic and military uses of cartography. Much has been written on the role of topographic maps in warfare, for example, from the Anglo-Boer War at the beginning of the century to the Gulf War in its closing decade. A comparable body of published research also exists on thematic maps as items of propaganda and persuasion, from the First World War to the Cold War.[2]

Cartography has been so important for national and imperial conquest and control that one could be forgiven for regarding maps as inherently acquisitive, proprietorial and confrontational.[3] And yet, as the other chapters in this volume demonstrate, maps are immensely complex visual representations of changing geographical realities and require interpretation from multiple perspectives. This chapter considers how maps were used during the twentieth century to promote the still-unrealised dreams of pacifists, anti-war activists and advocates of non-violence.[4]

It is not always a straightforward matter to distinguish between maps that were devoted to peace and international harmony and maps that were designed to facilitate warfare and military conquest. The various campaigns to build a more peaceful and harmonious world have frequently involved distributing maps designed for precisely the opposite purposes as a tactical manoeuvre to expose and challenge the often secret plans of military strategists and government officials. Conversely, maps initially intended to promote the causes of international peace and cooperation have sometimes been appropriated and transformed by national and imperial interests to reinforce more confrontational geopolitical and military objectives. In other words, war maps have sometimes become peace maps, and vice versa.

Claiming the earth

Cartographic images have been used to promote pacifism and internationalism for as long as these ideals have been articulated, though the number and variety of 'peace maps' increased significantly in the opening years of the twentieth century. As the globe seemed the 'natural' scale at which these ideals could be most effectively presented, the early twentieth century was an important moment in the 'cartographic genealogy' of the Earth as a symbol of unity and peace.[5]

The political symbolism of the globe was extensively deployed by a now forgotten group of early twentieth-century writers who sought to liberate the science of geography from its national, imperial and military associations, and relaunch the discipline as an overtly pacifist investigation of the Earth as a common resource, using ideas that foreshadow the arguments of the modern environmental movement. Some of these writers, notably the French geographer Élisée Reclus (1830–1905), espoused anarchist beliefs and advocated the radical overthrow of capitalism, nationalism and imperialism. Reclus and other geo-anarchist writers, including the Russians Peter Kropotkin (1842–1921) and Léon Metchnikoff (1838–88), saw cartography as a visual education that could promote international peace and a harmonious relationship between humanity and the natural world.

These ideas were exemplified by Reclus's ambitious scheme to construct a 'Great Globe' for the 1900 Paris Exposition Universelle (ill. 40), inspired in part by the success of James Wyld's more modestly proportioned structure, vividly described by *Punch* magazine as a 'geographical globule' that attracted huge crowds for more than a decade to London's Leicester Square after its unveiling during the 1851 Great Exhibition. The same ideas of pacifism and internationalism also motivated the Scottish socialist and polymath Patrick Geddes, who dreamed of establishing a National Institute of Geography in which he imagined two giant globes – one terrestrial, the other celestial – nestling either side of a new 'Outlook Tower', similar to the structure he had previously developed in Edinburgh.[6]

Although these delightful schemes progressed no further than the drawing board, their advocates proved relentlessly productive authors – particularly Reclus, who is best known for his nineteen-volume atlas and encyclopedia, *La Nouvelle Géographie Universelle: la Terre et les Hommes* (1875–94), translated into English under the title *The Earth and Its Inhabitants* (1882–95). This highly successful work presented a geographical portrait of the world as a single, unified space whose inhabitants could ultimately free themselves from the injustices and constraints of modern nations and empires, a surprisingly radical message for the largely middle-class readership who could afford these lavishly illustrated tomes. The same themes were explored in Reclus's powerful final work, *L'Homme et la Terre* (1905–8), the frontispiece of which presents a perfect early twentieth-century image of the Earth as home to a common humanity (ill. 41).

Despite the success of early twentieth-century radical geographers, their utopian convictions won only limited public support prior to the First World War. For those who

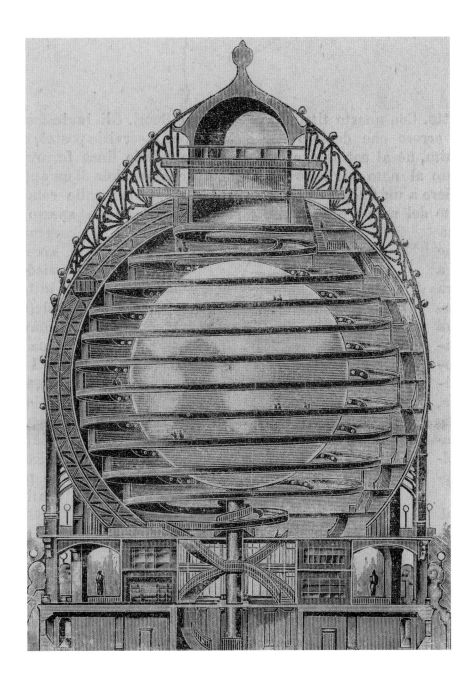

survived the killing fields in Flanders and Picardy, however, pacifism and internationalism seemed more necessary than ever if the 'Great War' was to become, as H. G. Wells had predicted at its outset, 'the war that will end war'.[7] The post-1918 peace movement, by no means limited to those on the left of the political spectrum, still revered the unitary globe as an inspirational symbol of hope and international reconciliation.

Throughout the interwar years, the undifferentiated terraqueous globe became a familiar architectural motif on public and commercial buildings and a popular logo for letterheads and flags of internationally minded organisations, agencies and corporations, including the new League of Nations, whose Geneva headquarters bristled with global imagery and iconography (ill. 42). Thematic world maps also featured regularly in the

ÉLISÉE RECLUS

L'HOMME ET LA TERRE

GEOGRAPHIE

HISTOIRE

PARIS
LIBRAIRIE UNIVERSELLE
33 RUE DE PROVENCE

DELOCHE SC
I.

41 (left)
Frontispiece to *L'Homme et la Terre*
by Élisée Reclus. Paris, 1905-8.
10007.m

42 (above)
League of Nations advisory
committee on traffic in opium,
with world map on the wall, 1938

League's publicity materials, issued to convince a still-uncertain public, within and beyond the member states, that the organisation was a worthwhile and effective venture (ill. 43).

Although the League had no specialist mapmaking facilities, and was therefore unable to challenge national (and often military-controlled) cartographic agencies, it eventually assembled an impressive specialist map library, bankrolled (as was the rest of the League's library) by the American oil billionaire John D. Rockefeller, Sr (1839–1937). The refusal of the United States to join the League opened a space for the Rockefeller family and other immensely wealthy American philanthropists, including the Scottish-born steel magnate Andrew Carnegie (1835–1919), to develop private foundations to promote the causes of international peace and reconciliation around the world, within the context of an American economic and political agenda. The Carnegie Endowment for International Peace, launched in 1910 with $10 million in high-yield mortgage bonds, was designed specifically to 'hasten the abolition of international war, the foulest blot upon our civilization', and concentrated initially on public education, particularly through its library-building programme. The Rockefeller Foundation, established three years later to advance 'the well-being of humanity throughout the world', reflected the Christian beliefs of Frederick Taylor Gates (1853–1929), the Rockefeller family's philanthropic advisor, and focused especially on medicine and health. The Rockefeller Foundation, in particular, deployed various images of the globe in its early logos (ill. 44).

SOCIAL QUESTIONS

TRAFFIC IN WOMEN AND CHILDREN	" Members will entrust the League with the general supervision over the execution of agreements with regard to the traffic in women and children and the traffic in opium and other dangerous drugs. "	PROTECTION OF WOMEN AND CHILDREN IN THE NEAR EAST
TRAFFIC IN OPIUM AND OTHER DANGEROUS DRUGS		CHILD WELFARE
OBSCENE PUBLICATIONS	(The Covenant.)	ARMENIAN REFUGEE SETTLEMENT SCHEME

Number of States signing the Protocol bringing the Convention into force.

1920

Before the League 6 — 1912, 1920
Under the League 45 — 1929

ANTI-OPIUM ACTIVITIES

1. The Hague Opium Convention 1912.
2. Creation of Advisory Committee exercising continuous supervision over execution of international obligations, and co-ordinating efforts of Governments.
3. Research for first time into world's medical requirements.
4. Two International Opium Conferences resulting in Opium Agreement and Opium Convention of 1925.
5. Permanent Central Opium Board established in 1928, on the coming into force of the Convention.
6. Enquiry in Persia on the possibilities of replacing the poppy by other crops.
7. Enquiry into control of opium-smoking in the Far East, 1929-1930.

TRAFFIC IN WOMEN AND CHILDREN

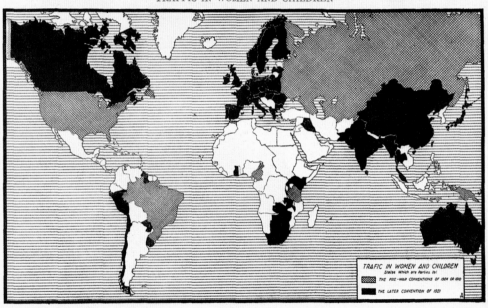

States which are Parties to the Conventions of 1904-10 and 1921.

1. International Conference summoned by the League, 1921, which resulted in :
(*a*) An International Convention to which 32 States, 26 British Mandated Territories and Crown Colonies, and the Colonies of the Netherlands and of Italy, are parties.
(*b*) An Advisory Committee, meeting first in 1922 and regularly once a year.
2. Enquiry in 28 countries by special body of experts into the extent of the traffic, 1924.

OBSCENE PUBLICATIONS

The International Conference summoned by the League in 1923 adopted a Convention to which 30 States, 41 British Crown Colonies, Protectorates and Mandated Territories, and the Colonies of the Netherlands, are parties.
Continuous supervision by the Traffic in Women and Children Committee.

43 (left)
'Traffic in Women and Children',
from *League of Nations: A Pictorial
Survey*, League of Nations, 1929

44 (above, left)
Logo of the Rockefeller
Foundation, 1913

45 (above, right)
The flag of the United
Nations, 1946

The larger and better resourced international organisations that emerged after the Second World War, including the United Nations (UN), were no less committed to the globe as the most powerful symbolic representation of the ideals that they represented and enforced. The UN's official emblem shows a map of the world on the azimuthal equidistant projection centred on the North Pole – though extending sixty degrees south – with the image surrounded by a wreath of crossed olive branches (ill. 45). This was intended to be an instantly recognisable 'aspirational symbol' that would encapsulate 'the hopes and dreams of people the world over for peace and unity'. The emblem was devised in San Francisco during the 1945 UN Conference on International Organization by a team of designers who had been commissioned by US Secretary of State Edward Stettinius, Jr to produce a distinctive lapel pin for the conference delegates.

The team was led by the architects Oliver Lincoln Lundquist (1916–2008) and Donal McLaughlin (1907–2003), both of whom had worked during the Second World War for the Office of Strategic Services (OSS), the forerunner of the Central Intelligence Agency (CIA). McLaughlin's contribution proved decisive, and he added the UN emblem to his other important commissions, including the interior of Tiffany's flagship store in Manhattan and the Nuremberg courtroom where leading Nazis were prosecuted after the Second World War.

The advent of the 'space age', although inspired by the Cold War competition between the USA and the USSR, generated a new wave of enthusiasm for global imagery and symbolism. The Unisphere, a giant stainless steel globe erected in the borough of Queens as the centrepiece of the 1964–5 New York World's Fair, was a particularly dramatic example (see ill. 39). This huge structure, forty-three metres high, was intended to represent the theme of 'peace through understanding', and specifically 'man's achievement in shrinking the globe', the latter symbolised by the three orbital rings that enveloped the structure in honour of the Russian cosmonaut Yuri Gagarin (the first man in space in 1961), the American astronaut John Glenn (the first US citizen to orbit the Earth in 1962) and the communications satellite Telstar (also launched in 1962).

The Unisphere's attempt to reimagine the globe as a symbol of international scientific cooperation in the context of the space age, in defiance of the confrontational geopolitics that inspired these costly space programmes, underlines how the exploration of outer space reinforced the idea of the Earth as a unifying symbol, capable of stirring the deepest human emotions. The first photographs of the Earth taken from outer space (from the famous *Earthrise* photograph from the Apollo 8 spacecraft as it circumnavigated the moon in 1968 to the stunning image of the entire unshadowed Earth during the Apollo 17 mission in 1972) had enormous cultural and political impact, especially for the environmental movement and the many non-governmental organisations, charitable foundations and humanitarian campaigns that used these 'one-world' images.[8]

The giant Live Aid banners displayed at London's Wembley Stadium and Philadelphia's John F. Kennedy Stadium for the charity rock concerts in the summer of 1985, each showing maps of the African continent set in an unmistakeably cartographic graticule, reaffirmed the emotional power of 'logo-cartography' for late twentieth-century humanitarian campaigns, in this case to raise money for the victims of the terrible famines that struck down an entire generation in East Africa (ill. 46). The less well-known 'Globe of Peace', ten metres in diameter and constructed as a labour of love

46

The Live Aid logo, emblazoned on banners decorating the stage at Wembley Stadium on 13 July 1985

by Orfeo Bartolucci in the hills outside the Italian village of Apecchio, near Pesaro, during the 1980s, had more modest consciousness-raising objectives, but remains an eloquent statement nevertheless (ill. 47).

If these late twentieth-century examples underline the enduring power of the globe as a symbol of internationalism and pacifism, it must also be acknowledged that the same image has been widely deployed for less selfless reasons by businesses and commercial interests anxious to promote their international ambitions, particularly in the United States, where film corporations have been especially keen on global imagery.

Universal Pictures, the first of the great Hollywood studios, has used a globe logo from its inception in 1912, and various spinning Earths have featured since the 1920s in the opening sequences of Universal films. The first example depicted a delightful little plane circumnavigating the Earth as if suspended on a wire. The latest incarnation transports the viewer on a sweeping, computer-generated visual journey across the Earth's surface while the company name, in shining gold letters, emerges from beyond the horizon to settle in front of a gently revolving globe as viewed from a satellite passing over at night, the sequence accompanied by a soaring musical score. RKO (Radio-Keith-Orpheum) Pictures, established in 1928, also had a famous publicity image showing a giant pulsating radio mast straddling the Earth's surface above the North

47
Orfeo Bartolucci, *Globo Terracqueo della Pace* (Globe of Peace), Apecchio, Italy. Completed in 1988

Pole. The British Broadcasting Corporation (BBC), established in 1922, adopted its own, less self-confident version of the spinning Earth for television broadcasts from 1963, an image that evolved into the famous 'mirror' sequence for the first colour transmission in 1969, and the single, high-resolution computer-generated image used from the early 1990s (ill. 48).

In recent years, military organisations concerned primarily with security and defence have also adopted the image of the globe to reflect their 'peace-keeping' roles. The 2003 renaming of the US National Imagery and Mapping Agency (NIMA) as the National Geospatial-Intelligence Agency (NGA) provides an interesting illustration. The NGA collects and analyses satellite-derived geospatial intelligence for all departments of the US federal government from its enormous headquarters in Fort Belvoir, on the outskirts of Washington DC, the third-largest government building in the United States. Prior to 2003, NIMA's logo confidently asserted its military role in preserving Pax Americana by reference to eagles and broken arrows. Since 2003, NGA's logo seems designed to mask its geo-strategic and surveillance activities behind a comforting 'Earthrise' globe showing the entire American continent emerging into the sunlight, watched over by a satellite 'star' twinkling, as if on a Christmas card, in the vastness of space. In place of NIMA's bewildering Latin motto *Tempestivum verum definitum* (Time defined truth), we now have NGA's 'Know the Earth, show the way' (ill. 49).

Projecting peace

Alongside these global representations, there have been several more technical cartographic projects inspired, at least initially, by the same internationalist and even pacifist objectives, though these goals have often been compromised by traditional geopolitical rivalries. The project to compile an International Map of the War (IMW) on a 1:1,000,000 scale is an excellent example. This was first proposed in 1891 by the German geographer Albrecht Penck (1858–1945). The idea was to persuade the leading national

48 (above left)
The BBC's 'spinning earth logo' from 1963

49 (above right)
The National Imagery and Mapping Agency was renamed in 1993 as the National Geospatial Intelligence Agency, as reflected in their original and later logos.

mapping agencies to cooperate in preparing a new map series, compiled from the existing cartographic archive and based on the Greenwich meridian, the metric system of measurement and a common set of symbols and conventions. The objective was a unified, internationally agreed global map series to mark the beginning of the twentieth century. According to Penck, the IMW would simultaneously summarise and transcend the complex archive of national and imperial maps that had developed piecemeal over the previous centuries. It would serve as a symbol of peaceful scientific collaboration.[9]

International conferences were convened, backed by national governments, in London in 1909 and in Paris in 1913 to establish more precise guidelines, though with limited success. The withdrawal of the United States before the Paris conference, and the outbreak of the First World War a few months later, brought formal proceedings to a shuddering halt. The Royal Geographical Society (RGS) in London continued to produce IMW-style 1:1,000,000 map sheets throughout the war, but as these were prepared under the auspices of the War Office to assist British military and geopolitical preparations for the post-war negotiations, they were not remotely consistent with the project's original internationalist motivations. Indeed, the wartime RGS map sheets were extensively used to prepare ethnographic maps before and during the Paris Peace Conferences to legitimise the dismemberment of the Austro-Hungarian and Ottoman empires.[10]

The IMW project was revived on something like its original terms after 1918, with the backing of the League of Nations, though the initial enthusiasm was never fully recaptured, partly because Penck, angered by Germany's treatment after the First World War, rejected his pre-war internationalism in favour of a more conservative and nationalist geopolitics. Several countries also embarked on their own million-scale mapping projects for conventional national and imperial purposes, while paying lip-service to the IMW's original ideals.

The magnificent 100-plus sheets of the 1:1,000,000 *Map of Hispanic America*, prepared at great expense during the 1920s and 1930s by cartographers in the American Geographical Society in New York, is perhaps the best example of how the IMW, originally intended to promote international peace and cooperation, eventually served to support traditional national and imperial objectives. The *Map of Hispanic America* stands as a monument to US cultural imperialism in Latin America, even though the cartographers involved in the project insisted they were motivated solely by benign scientific interests, repeatedly emphasising how the Hispanic map could help to resolve border disputes across the region (ill. 50).[11] The official IMW project continued throughout this period, however, and more than 200 map sheets were compiled by forty-four national cartographic agencies during the 1920s, though only twenty-one conformed exactly to the original specifications agreed in Paris in 1913 (ill. 51).

The IMW project continued after the Second World War, under the auspices of the United Nations from the early 1950s, though it gradually became clear that its original cultural and geopolitical rationale was no longer relevant in the context of the Cold War and given the developments in other forms of mapping at that scale using aerial reconnaissance.[12] Communist cartographers in the Soviet Union and allied states in

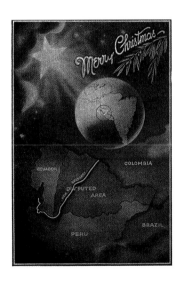

50 (above)
George McBride, Christmas card, 1942. The card illustrates the Equadorian-Peruvian border dispute. American Geographical Society

51 (next page)
International Map of the World sheet, covering parts of India, East Pakistan (Bangladesh), Bhutan, China, Sikkim and Nepal (Sheet NG-45). *Survey of Pakistan*, 1963. Maps 920.(294)

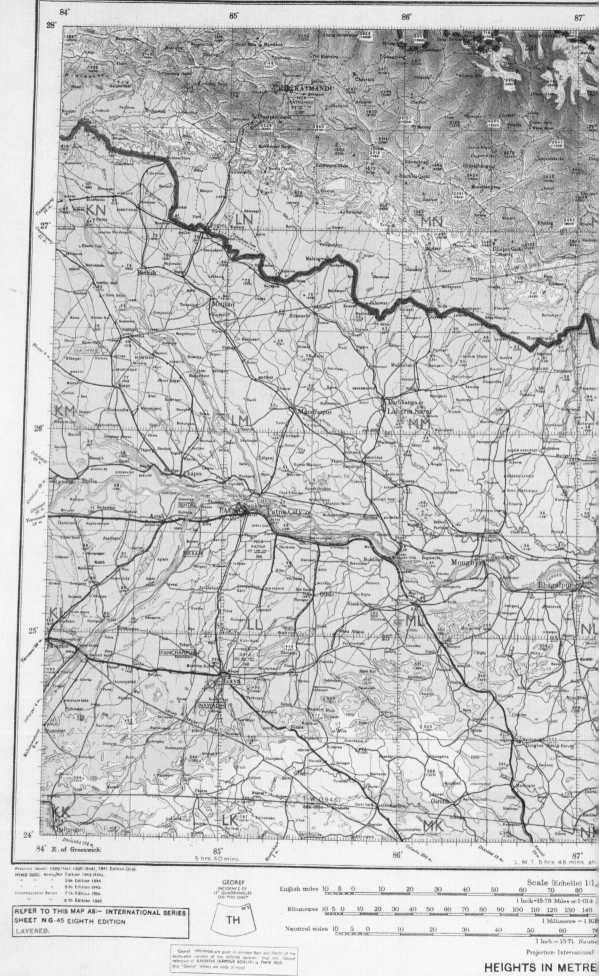

Previous issues: 1920 (1st), 1930 (2nd), 1941 Edition (3rd).
HIND 5000, Army/Air Edition 1943 (4th).
                    5th Edition 1944.
                    6th Edition 1945.
International Series   7th Edition 1956.
                    8th Edition 1965.

**REFER TO THIS MAP AS:— INTERNATIONAL SERIES
SHEET N G-45 EIGHTH EDITION**
LAYERED.

GEOREF
INCIDENCE OF
15° QUADRANGLES
ON THIS SHEET

TH

Georef references are given in minutes East and North of the
south-west corners of the lettered squares; thus the Georef
reference of RASHAHI (RAMPUR BOALIA) is TMPK 3623.
(For 'Georef' letters see body of map).

Scale (Echelle) 1:1,

English miles 10 5 0   10   20   30   40   50   60   70   80
                                                                    1 Inch =15·78 Miles or 1·014 ·

Kilometres 10 5 0 10 20 30 40 50 60 70 80 90 100 110 120 130 140
                                                                    1 Millimetre = 1 Kil

Nautical miles 10 5 0   10   20   30   40   50   60   76
                                                                    1 Inch = 13·71 Nautic

Projection: International

HEIGHTS IN METRE

Certain main spot heights are indicated by al

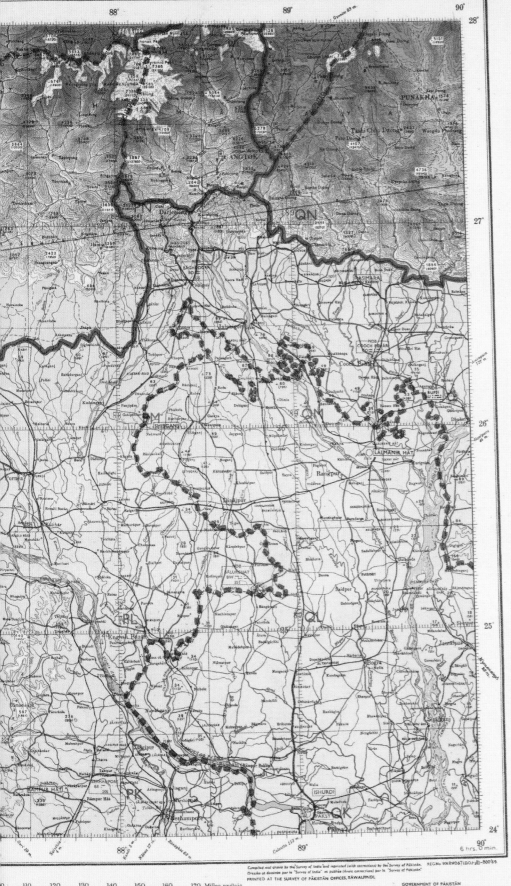

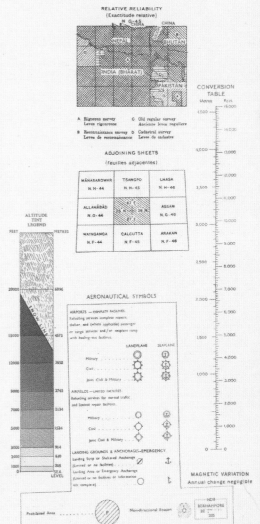

Compiled and drawn by the Survey of India and reprinted (with corrections) by the Survey of Pakistan. REG.No.906 RWD 67 (D.O.I-p)-500'69.
PRINTED AT THE SURVEY OF PAKISTAN OFFICES, RAWALPINDI.

FEET

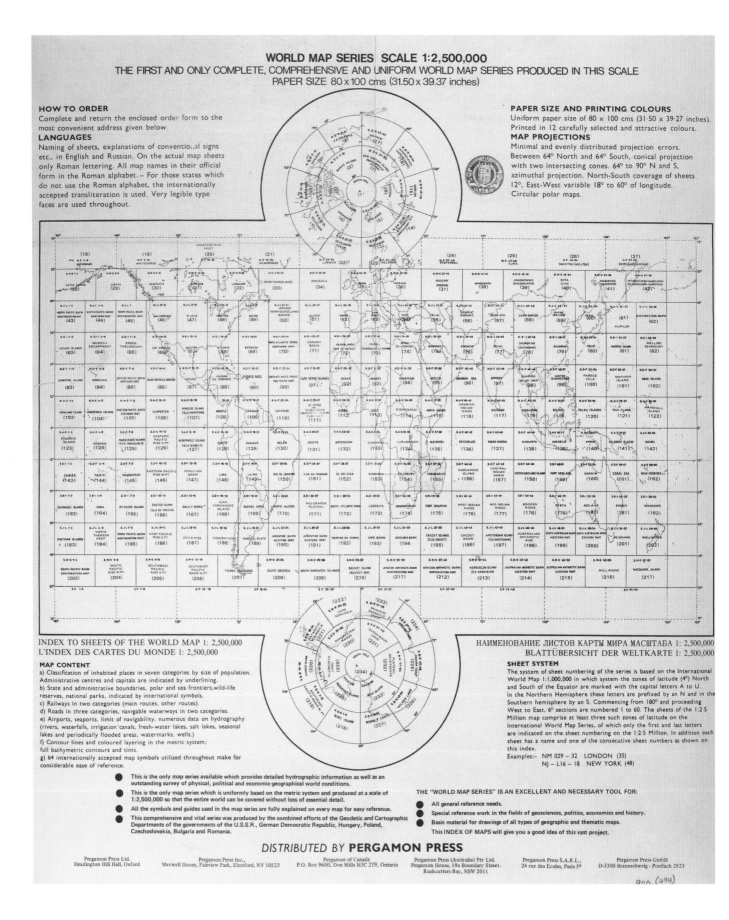

WORLD MAP SERIES SCALE 1:2,500,000
THE FIRST AND ONLY COMPLETE, COMPREHENSIVE AND UNIFORM WORLD MAP SERIES PRODUCED IN THIS SCALE
PAPER SIZE 80 x 100 cms (31.50 x 39.37 inches)

HOW TO ORDER
Complete and return the enclosed order form to the most convenient address given below

LANGUAGES
Naming of sheets, explanations of conventional signs etc., in English and Russian. On the actual map sheets only Roman lettering. All map names in their official form in the Roman alphabet. – For those states which do not use the Roman alphabet, the internationally accepted transliteration is used. Very legible type faces are used throughout.

PAPER SIZE AND PRINTING COLOURS
Uniform paper size of 80 x 100 cms (31.50 x 39.27 inches). Printed in 12 carefully selected and attractive colours.

MAP PROJECTIONS
Minimal and evenly distributed projection errors. Between 64° North and 64° South, conical projection with two intersecting cones. 64° to 90° N and S, azimuthal projection. North-South coverage of sheets 12°, East-West variable 18° to 60° of longitude. Circular polar maps.

INDEX TO SHEETS OF THE WORLD MAP 1: 2,500,000
L'INDEX DES CARTES DU MONDE 1: 2,500,000

MAP CONTENT
a) Classification of inhabited places in seven categories by size of population. Administrative centres and capitals are indicated by underlining.
b) State and administrative boundaries, polar and sea frontiers, wild-life reserves, national parks, indicated by international symbols.
c) Railways in two categories (main routes, other routes).
d) Roads in three categories, navigable waterways in two categories.
e) Airports, seaports, limit of navigability, numerous data on hydrography (rivers, waterfalls, irrigation canals, fresh-water lakes, salt lakes, seasonal lakes and periodically flooded areas, watermarks, wells).
f) Contour lines and coloured layering in the metric system; full bathymetric contours and tints.
g) 64 internationally accepted map symbols utilized throughout make for considerable ease of reference.

● This is the only map series available which provides detailed hydrographic information as well as an outstanding survey of physical, political and economic-geographical world conditions.
● This is the only map series which is uniformly based on the metric system and produced at a scale of 1:2,500,000 so that the entire world can be covered without loss of essential detail.
● All the symbols and guides used in the map series are fully explained on every map for easy reference.
● This comprehensive and vital series was produced by the combined efforts of the Geodetic and Cartographic Departments of the governments of the U.S.S.R., German Democratic Republic, Hungary, Poland, Czechoslovakia, Bulgaria and Romania.

НАИМЕНОВАНИЕ ЛИСТОВ КАРТЫ МИРА МАСШТАБА 1: 2,500,000
BLATTÜBERSICHT DER WELTKARTE 1: 2,500,000

SHEET SYSTEM
The system of sheet numbering of the series is based on the International World Map 1:1,000,000 in which system the zones of latitude (4°) North and South of the Equator are marked with the capital letters A to U. In the Northern Hemisphere these letters are prefixed by an N and in the Southern hemisphere by an S. Commencing from 180° and proceeding West to East, 6° sections are numbered 1 to 60. The sheets of the 1:2·5 Million map comprise at least three such zones of latitude on the International World Map Series, of which only the first and last letters are indicated on the sheet numbering on the 1:2·5 Million. In addition each sheet has a name and one of the consecutive sheet numbers as shown on this index.
Examples:– NM 029 – 32 LONDON (35)
              NJ – L16 – 18 NEW YORK (48)

THE "WORLD MAP SERIES" IS AN EXCELLENT AND NECESSARY TOOL FOR:
● All general reference needs.
● Special reference work in the fields of geosciences, politics, economics and history.
● Basic material for drawings of all types of geographic and thematic maps.
This INDEX OF MAPS will give you a good idea of this vast project.

DISTRIBUTED BY **PERGAMON PRESS**

Pergamon Press Ltd.
Headington Hill Hall, Oxford

Pergamon Press Inc.,
Maxwell House, Fairview Park, Elmsford, NY 10523

Pergamon of Canada
P.O. Box 9600, Don Mills M3C 2T9, Ontario

Pergamon Press (Australia) Pty Ltd.
Pergamon House, 19a Boundary Street,
Rushcutters Bay, NSW 2011

Pergamon Press S.A.R.L.,
24 rue des Ecoles, Paris 5ᵉ

Pergamon Press GmbH
D-3300 Braunschweig · Postfach 2923

900 (494)

Eastern Europe completed their own version of the international map, the Karta Mira, on the less politically contentious scale of 1:2,500,000 during the 1960s and 1970s. The Karta Mira sheets were often made available gratis in Western countries, a fascinating exercise in communist propaganda facilitated in part by the British publisher and newspaper proprietor Robert Maxwell (ill. 52). The IMW project gradually fizzled out in the 1970s, to the evident relief of some cartographers. According to Arthur Robinson, the redoubtable American geographer writing in the mid-1960s, the IMW had become a pointless irrelevance, a form of 'cartographic wallpaper'.[13]

Others remained sympathetic to the IMW's original inspiration while regretting the project's inability to transcend national and imperial rivalries. In 1974 the German filmmaker and historian Arno Peters (1916–2002) took up the challenge of creating a new cartographic image of the world in the name of international peace. The Peters proposal was a radical solution to the age-old problem of 'projecting' the three-dimensional globe onto a two-dimensional map and took the form of a new equal-area projection of the Earth's land masses, a single 'world-picture' which Peters insisted was more conducive to the spirit of internationalism that had originally inspired the IMW (ill. 53).

The Peters projection, as it was soon known, was initially devised to facilitate an educational programme which its author had been promoting since the 1950s under the title 'Synchronoptische Weltgeschichte' (Synchronoptic World History). This was a form of global history in which all regions of the world were afforded the same consideration, supposedly to reveal previously overlooked geographical relationships and interconnections.[14] Peters claimed that his vision of the world's geography challenged the Eurocentric, nationalist and imperialist connotations of earlier map projections. His main target was the Mercator projection, formulated in 1569 by the Flemish mapmaker Gerardus Mercator (1512–94) and the basis of most European and North American atlases and wall maps from the seventeenth century onwards. To accommodate the unequal distribution of the world's land masses in the northern and southern hemispheres, the Mercator projection distorted the relative sizes and shapes of territories at higher latitudes and frequently depicted the equator below the middle of the map, further reinforcing the visual significance of the northern hemisphere.

According to Peters, the Mercator projection was hopelessly anachronistic and betrayed the imperial assumptions and cultural arrogance of the European cartographers who had used it continuously since its inception. This was an image of the world that asserted Europe's global centrality at the expense of tropical and southern regions. The Peters projection, by contrast, placed Africa squarely in the centre of the map and emphasised the increasingly populous regions of the southern hemisphere. This was a new projection for a new age, claimed Peters: 'an egalitarian world map which alone can demonstrate the parity of peoples of the globe'.

The Peters projection generated a lengthy and acrimonious controversy. Several professional cartographers dismissed its author as a politically motivated self-publicist with no background or training in the field. The aforementioned Arthur Robinson mocked the elongated land masses on the Peters projection which he likened to 'wet, ragged, long winter underwear hung out to dry'. This was in any case not a new visualisation, insisted

52 (left)
Index to the *Karta Mira*
(Map of the World), 1964-.
Maps 920.(494)

53 (next page)
Map of the World: Peters Projection.
Oxfam Publications, *c.* 1989.
Maps X.1199

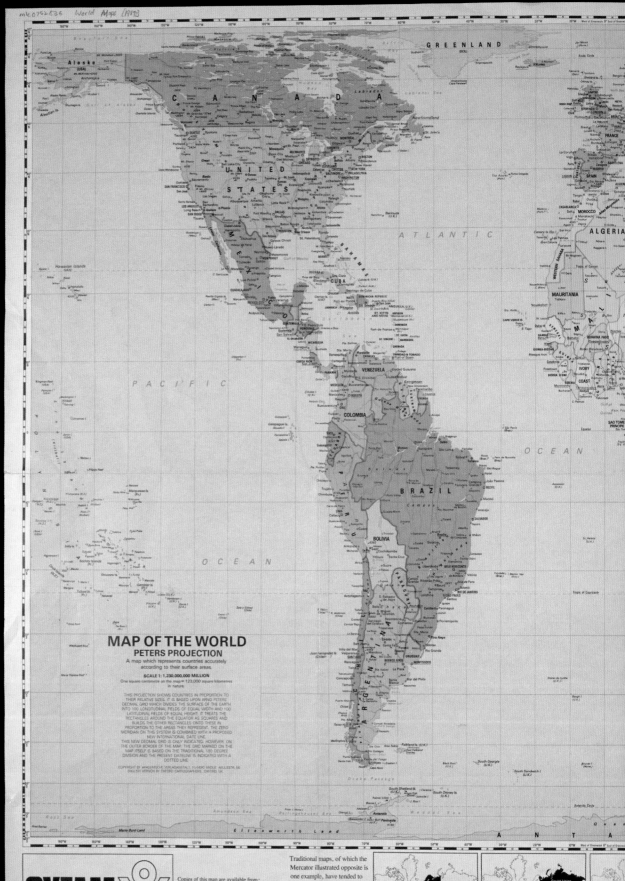

MAP OF THE WORLD
PETERS PROJECTION
A map which represents countries accurately
according to their surface areas.

SCALE 1: 1,230,000,000 MILLION
One square centimetre on the map = 123,000 square kilometres
in nature.

THIS PROJECTION SHOWS COUNTRIES IN PROPORTION TO
THEIR RELATIVE SIZES. IT IS BASED UPON ARNO PETERS'
DECIMAL GRID WHICH DIVIDES THE SURFACE OF THE EARTH
INTO 100 LONGITUDINAL FIELDS OF EQUAL WIDTH AND 100
LATITUDINAL FIELDS OF EQUAL HEIGHT. IT TREATS THE
RECTANGLES AROUND THE EQUATOR AS SQUARES AND
BUILDS THE OTHER RECTANGLES ONTO THESE IN
PROPORTION TO THE AREAS THEY REPRESENT. THE ZERO
MERIDIAN ON THIS SYSTEM IS COMBINED WITH A PROPOSED
NEW INTERNATIONAL DATE LINE.
THIS NEW DECIMAL GRID IS ONLY INDICATED, HOWEVER ON
THE OUTER BORDER OF THE MAP. THE GRID MARKED ON THE
MAP ITSELF IS BASED ON THE TRADITIONAL 180 DEGREE
DIVISION AND THE PRESENT DATELINE IS INDICATED WITH A
DOTTED LINE.

COPYRIGHT BY AKADEMISCHE VERLAGANSTALT, FL/SBERG VADUZ, HELLSTR. 56.
ENGLISH VERSION BY OXFORD CARTOGRAPHERS, OXFORD, UK.

OXFAM

Copies of this map are available from:

OXFAM Publications
274 Banbury Road
Oxford OX2 7DZ
Tel (0865) 56777

ISBN 0855981296

Traditional maps, of which the
Mercator illustrated opposite is
one example, have tended to
show countries incorrectly in
proportion to one another, to
the advantage of the European
colonial powers, while the
southern continents (Africa,
South America, Australia) are
shown far too small.

Europe, with its 9.7 million sq km appears to be
larger than South America which is 17.8 million
sq km, in fact twice the size of Europe.

The Soviet Union appears to be considerably
larger than Africa which is 30 million sq km. In
fact it is smaller. (actually 22.4 million sq km).

Scandinavia, 1.1
large as India wh
is three times as

MAPS X.1199.

PETERS PROJECTION

This new map, the work of the German historian Arno Peters, provides a helpful corrective to the distortions of traditional maps. While the Peters Map is superior in its portrayal of proportions and sizes, its importance goes far beyond questions of cartographic accuracy. No less than our world view is at stake.

...ms to be as ... In fact India ...via.

Greenland with 2.1 million sq km appears larger than China when in fact China with 9.5 million sq km is about four times larger.

Consider the characteristics of the Peters Map:

EQUAL AREA. This new map shows all areas — whether countries, continents or oceans — according to their actual size. Accurate comparisons become possible.

EQUAL AXIS. All North-South lines run vertically on this map. Thus, geographic points can be seen in their precise directional relationship — northwest, southeast, northeast or southwest.

EQUAL POSITIONS. All East-West lines run parallel. Thus the relationship of any point on the map to its distance from the equator or to the angle of the sun can easily be determined.

FAIRNESS TO ALL PEOPLES. By setting all countries in their true size and location, this map allows each one its actual position in the world.

In this complex and interdependent world in which nations now live, the peoples of the world deserve the most accurate possible portrayal of their world. The Peters Map is that map for our day.

Maps X. 1199

erely a minor reworking of a projection suggested more than a century
Gall (1808–95), a Scottish clergyman and astronomer, whose objective
movements of stars across the Earth's surface. Others noted that a
n had been proposed just a few years earlier by Trystan Edwards (1884–
architect, town planner and journalist, and hinted darkly that Peters
s own scheme, amid much fanfare, only after Edwards had died so as
of plagiarism.

hostile reception, the Peters projection (or the Gall–Peters projection
aphers insisted it should be called) was adopted by several charities,
n-governmental organisations that shared the explicitly internationalist,
objectives of its author. In 1977 the British charity Christian Aid
colour promotional poster featuring the original German version of
ction alongside a commentary on why this world-view was more in
values and objectives than the Mercator projection. The US-based
il of Churches also began to use the projection in its publicity. The
a Peters-projected world map for the dramatic and now iconic cover of
t of the Independent Commission on International Development Issues
South: A Program for Survival (generally known as the Brandt Report in
airman, guiding spirit and main author – Nobel Peace Prize winner and
Chancellor Willi Brandt) brought a new level of international familiarity
. The map continued to be used in later editions (ill. 54).
Report was hugely influential on international development debates,
n the United Nations, which also began to use the Peters projection
itically correct way to depict the world's geography. Newspapers and
icularly those on the left, made increasing use of the projection for their
as did the specialist international development and activist literature,
ford-based radical magazine New Internationalist, which regularly
lour world maps based on the Peters projection on its front covers. In
he United Nations Children's Emergency Fund, issued a Christmas appeal
he Peters projection, which was updated every year thereafter for other
blicity. To capitalise on his success, Peters published his own atlas of the
o demonstrate how his projection could be used for complex thematic

of the twentieth century, UNICEF had distributed more than 60 million
the Peters projection, and this now familiar world-view had appeared in
reports prepared by UNESCO and most leading international charities,
n. The appearance of the Peters projection in promotional literature
e Catholic Agency for Overseas Development (CAFOD) reflected the
on to adopt this global image. The story of the Peters projection even
episode of the cult US television series The West Wing in 2001.
to maps designed specifically to promote the cause of international
protest groups have occasionally used incriminating and even secret
to facilitate warfare as part of their campaigns. In an attempt to discredit

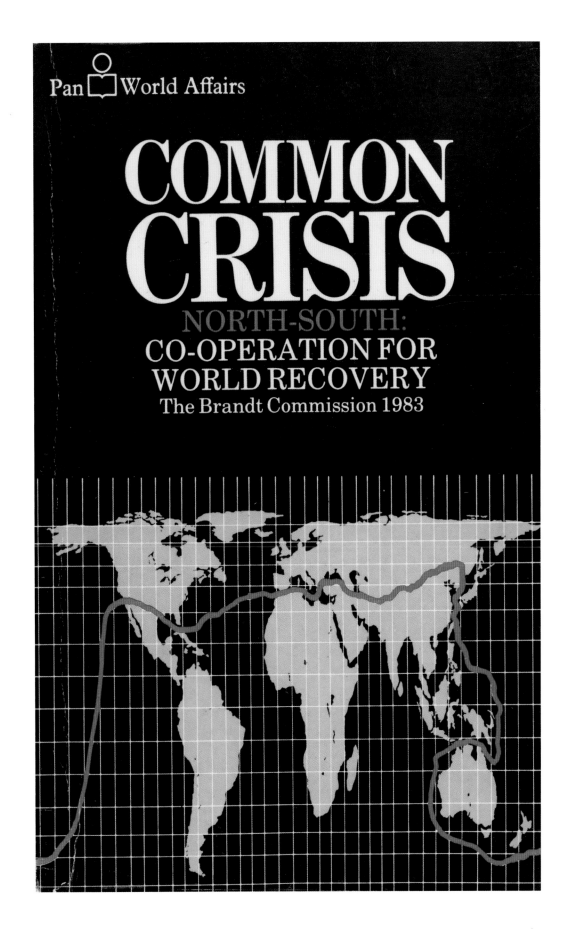

the doctrine of 'mutually assured destruction' and to prove that the British government was planning for a nuclear exchange, the Campaign for Nuclear Disarmament (CND) reprinted official maps and prepared its own cartography based on leaked official statistics, including the information acquired by the investigative journalist Duncan Campbell from the 'Square Leg' exercise undertaken by strategists in the Ministry of Defence in 1980. This exercise assessed the likely impact of a 250-megaton attack on the UK involving 131 nuclear weapons, a scenario subsequently revealed to be a massive underestimate. The results predicted the immediate death of 29 million people (53 per cent of the total population) and serious, life-threatening injuries for a further 7 million (12 per cent). Only around 19 million (39 per cent) were predicted to survive.[15] These and other chilling calculations were converted into a CND map showing the likely fallout from such an attack, and the data was further developed into a full-length book on Britain after a nuclear attack by professional cartographers sympathetic to the cause of unilateral nuclear disarmament (ills 55, 56).[16]

Six years later, the radical American geographer William Bunge produced a more international *Nuclear War Atlas* to make the same arguments. This compendium of innovative 'shock maps' was intended to convey the horrific consequences of thermonuclear war, based on the conviction that an otherwise complacent public needed to be jolted into action to 'protest and survive' (ill. 57).[17]

The British social historian E. P. Thompson, a leading advocate of international nuclear disarmament, also relied on shock maps during a famous confrontation with the US Secretary of Defense, Caspar Weinberger, at the Oxford Union, the University debating society, in 1984. The two men, and others, had been invited to debate the proposition that 'There is no moral difference between the foreign policies of the US and the USSR'. Thompson illustrated his impassioned remarks with maps depicting likely nuclear exchanges printed in two pamphlets – one prepared by Weinberger's own department in Washington, the other by the equivalent ministry in Moscow. According to Thompson, these 'two odious books', bound together, would 'make the most evil book known in the whole human record' – a dramatic claim, though it failed to convince a majority of the listening students, as 271 voted against the motion and 232 in support.[18]

Ideological 'map wars' of this kind have continued and intensified in recent years, despite the end of the Cold War. The emergence of the Internet has had a transformative impact, as twenty-first-century political campaigning is now largely conducted online and through social media. Opposing campaign groups in the Middle East, for example, now make particularly imaginative use of digital technology and interactive cartography to advance their arguments online – notably the Peace Now movement, which campaigns for a two-state solution to the Israel–Palestine conflict using astonishingly intricate maps (ill. 58).

The Institute for Economics and Peace (IEP), an independent think-tank established by the Australian technology entrepreneur Steve Killelea with offices in Sydney, New York and Mexico, is especially interesting this this regard. The IEP claims to be 'dedicated to shifting the world's focus to peace as a positive, achievable and tangible measure of human well-being and progress'. To advance this laudable objective, the IEP has been

55
Map of Operation Square Leg, originally created by CND with help from investigative journalist Duncan Campbell. The map shows ground and airburst weapons and the plumes of radioactive fallout. Duncan Campbell, *War Plan UK*, Burnett, 1982.
X.629/23118

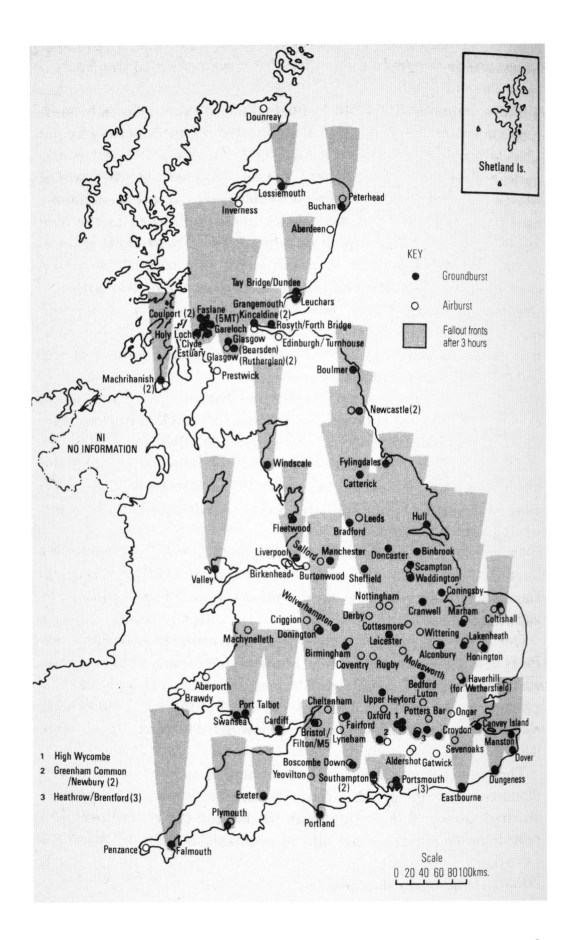

Dounreay

Lossiemouth
Inverness   Buchan Peterhead

Aberdeen

Tay Bridge/Dundee

Grangemouth/   Leuchars
Coulport (2) Faslane Kincaldine (2)
              (5MT)   Rosyth/Forth Bridge
Holy Loch Gareloch
\Clyde   Glasgow   Edinburgh/Turnhouse
Estuary   (Bearsden)
      Glasgow (Rutherglen)(2)
Machrihanish   Boulmer
(2)   Prestwick

                                      Newcastle(2)

NI
NO INFORMATION
                    Windscale   Fylingdales

                                   Catterick

                              Leeds   Hull
              Fleetwood   Bradford
                        Salford
              Liverpool Manchester Doncaster Binbrook
                                                Scampton
              Birkenhead Burtonwood Sheffield Waddington
Valley   Coningsby
                              Nottingham
                    Wolverhampton   Cranwell Marham
              Criggion   Derby   Coltishall
                          Donington Cottesmore Wittering Lakenheath
Machynelleth   Leicester   Honington
                          Birmingham   Alconbury
                      Coventry Rugby Molesworth
                                          Bedford   Haverhill
Aberporth   Luton   (for Wethersfield)
Brawdy
          Port Talbot   Cheltenham Upper Heyford   Ongar
Swansea   Cardiff   Oxford 1 Potters Bar   Canvey Island
                        Bristol/   Fairford 2   3 Croydon Manston
                        Filton/M5 Lyneham   Sevenoaks Dover
                                              Aldershot Gatwick
                  Boscombe Down   Dungeness
        Yeovilton Southampton   Portsmouth   Eastbourne
                        (2)   (3)
          Exeter

      Plymouth
            Portland

Penzance Falmouth

KEY

● Groundburst

○ Airburst

▨ Fallout fronts after 3 hours

Shetland Is.

Scale
0 20 40 60 80 100kms.

1 High Wycombe
2 Greenham Common /Newbury (2)
3 Heathrow/Brentford (3)

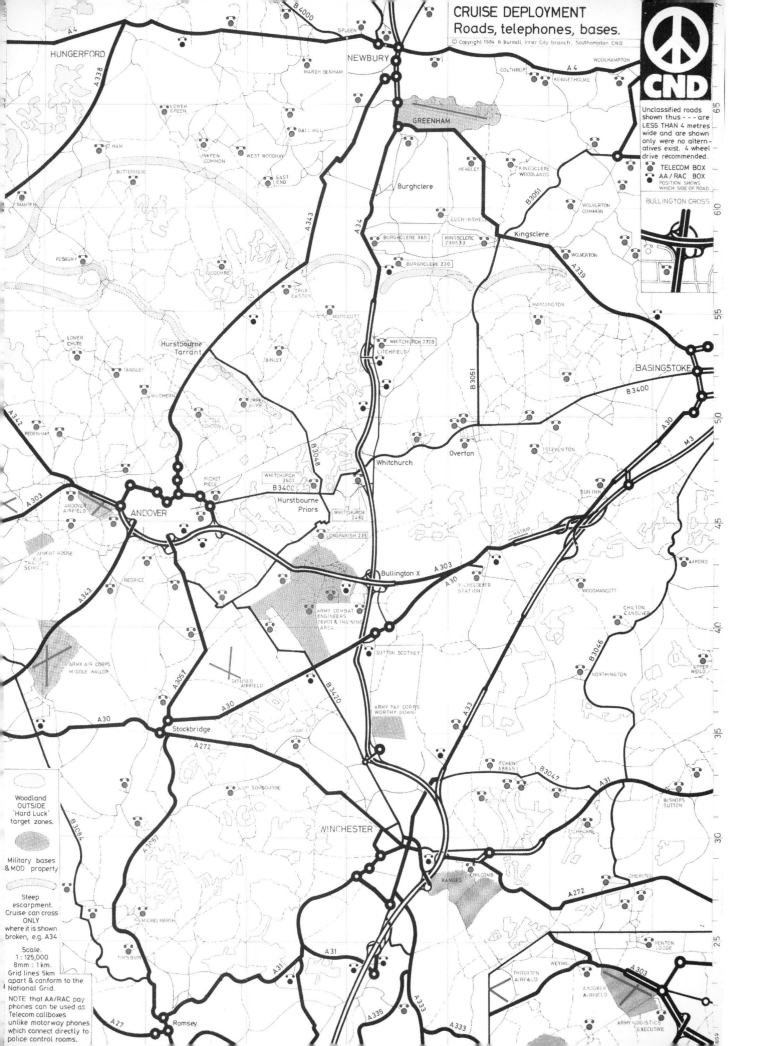

CRUISE DEPLOYMENT
Roads, telephones, bases.

© Copyright 1984 B Burnell, Inner City branch, Southampton CND

Unclassified roads shown thus - - - are LESS THAN 4 metres wide and are shown only were no alternatives exist. 4 wheel drive recommended.

⬤ TELECOM BOX
⬤ AA/RAC BOX
POSITION SHOWS WHICH SIDE OF ROAD.

BULLINGTON CROSS

Woodland OUTSIDE 'Hard Luck' target zones.

Military bases & MOD property.

Steep escarpment. Cruise can cross ONLY where it is shown broken, e.g. A34

Scale. 1 : 125,000
8mm : 1km.
Grid lines 5km apart & canform to the National Grid.

NOTE that AA/RAC pay phones can be used as Telecom callboxes unlike motorway phones which connect directly to police control rooms.

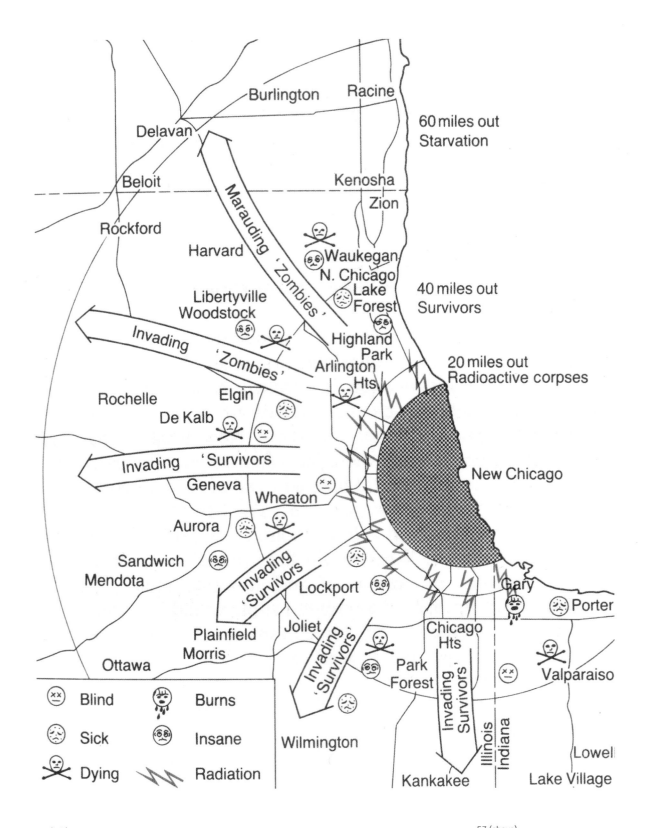

Burlington   Racine

Delavan

60 miles out
Starvation

Beloit   Kenosha

Zion

Rockford

Harvard   Waukegan
N. Chicago
Lake
Forest

40 miles out
Survivors

Marauding 'Zombies'

Libertyville
Woodstock

Invading 'Zombies'

Highland
Park

Arlington
Hts

20 miles out
Radioactive corpses

Rochelle   Elgin

De Kalb

Invading 'Survivors'

Geneva

Wheaton

New Chicago

Aurora

Sandwich
Mendota

Invading 'Survivors'

Lockport

Gary   Porter

Joliet

Plainfield
Morris

Invading 'Survivors'

Chicago
Hts

Invading 'Survivors'

Valparaiso

Ottawa

Park
Forest

Illinois
Indiana

Lowel

Wilmington

Kankakee   Lake Village

(XX) Blind   Burns

Sick   Insane

Dying   Radiation

56 (left)

Cruise Deployment: Roads, Telephones, Bases.
Campaign for Nuclear Disarmament map for use by
protestors disrupting US military manoeuvres, 1984.
Maps CC.5.a.416

57 (above)
William Bunge, 'New Chicago,' from
Nuclear War Atlas. Blackwell, 1988.
YC.1988.a.12507

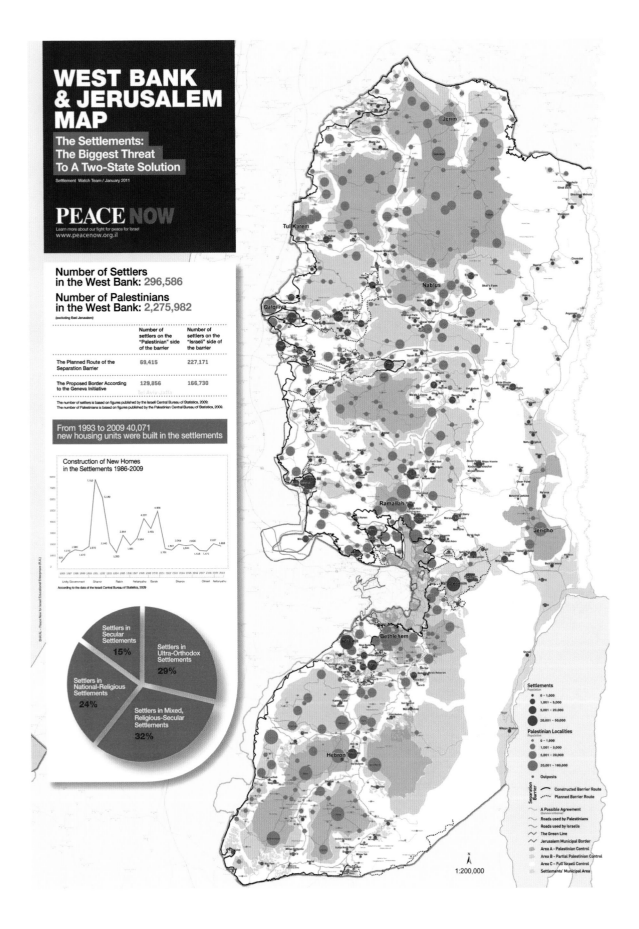

WEST BANK & JERUSALEM MAP

The Settlements: The Biggest Threat To A Two-State Solution

Settlement Watch Team / January 2011

PEACE NOW

Learn more about our fight for peace for Israel
www.peacenow.org.il

Number of Settlers in the West Bank: 296,586

Number of Palestinians in the West Bank: 2,275,982

(excluding East Jerusalem)

	Number of settlers on the "Palestinian" side of the barrier	Number of settlers on the "Israeli" side of the barrier
The Planned Route of the Separation Barrier	69,415	227,171
The Proposed Border According to the Geneva Initiative	129,856	166,730

The number of settlers is based on figures published by the Israeli Central Bureau of Statistics, 2009;
The number of Palestinians is based on figures published by the Palestinian Central Bureau of Statistics, 2000.

From 1993 to 2009 40,071 new housing units were built in the settlements

Construction of New Homes in the Settlements 1986-2009

According to the data of the Israeli Central Bureau of Statistics, 2009

Pie chart:
- Settlers in Secular Settlements 15%
- Settlers in Ultra-Orthodox Settlements 29%
- Settlers in National-Religious Settlements 24%
- Settlers in Mixed, Religious-Secular Settlements 32%

Settlements
Population
- 0 - 1,000
- 1,001 - 5,000
- 5,001 - 20,000
- 20,001 - 50,000

Palestinian Localities
Population
- 0 - 1,000
- 1,001 - 5,000
- 5,001 - 20,000
- 20,001 - 160,000
- Outposts

Separation Barrier
- Constructed Barrier Route
- Planned Barrier Route

- A Possible Agreement (Geneva Initiative)
- Roads used by Palestinians
- Roads used by Israelis
- The Green Line
- Jerusalem Municipal Border
- Area A – Palestinian Control
- Area B – Partial Palestinian Control
- Area C – Full Israeli Control
- Settlements' Municipal Area

1:200,000

'developing new conceptual frameworks to define peacefulness; providing metrics for measuring peace; and uncovering the relationships between business, peace and prosperity as well as promoting a better understanding of the cultural, economic and political factors that create peace' – notably through a Global Peace Index (GPI) which forms the basis of several IEP maps at various scales, from that of individual countries (USA, UK and Mexico) to the whole globe. These can be accessed through an IEP website, Vision of Humanity.[19]

The IEP website suggests that the peace campaign now uses maps in entirely new ways, though the otherwise conventional nature of these spatial representations, which would be instantly recognisable to Reclus or Penck, reaffirms a far deeper truth: that the concept of world peace ultimately relies on a geographical imagination.

58
Peace Now, *West Bank and Jerusalem Map*, 2011

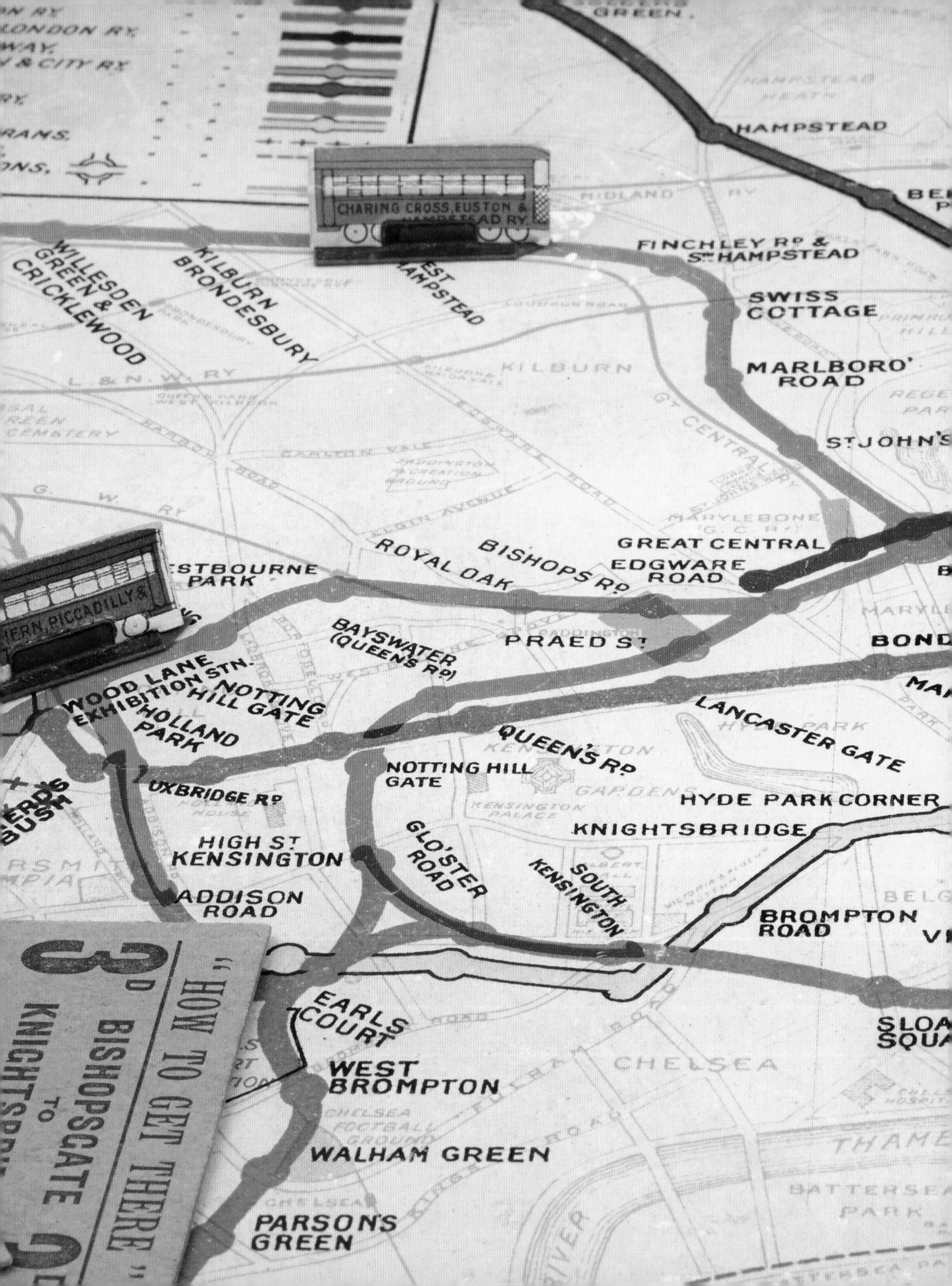

LONDON RY
WAY
& CITY RY.
RY.
RANS;
NS;

HAMPSTEAD
BE... P...

CHARING CROSS EUSTON & HAMPSTEAD RY.

FINCHLEY RD. & Sᵗ HAMPSTEAD
SWISS COTTAGE
MARLBORO' ROAD
St. JOHN'S

MIDLAND RY

EST HAMPSTEAD

WILLESDEN GREEN & CRICKLEWOOD
KILBURN BRONDESBURY

L & N W RY

KILBURN

GT CENTRAL RY

REGENTS PARK
MARYLEBONE (G.C.R.)
GREAT CENTRAL
EDGWARE ROAD

BISHOPS RD.
ROYAL OAK
ESTBOURNE PARK

G W RY

PRAED Sᵗ
BOND

MARYLE...

...HERN, PICCADILLY &

BAYSWATER (QUEENS RD.)

LANCASTER GATE
HYDE PARK

WOOD LANE EXHIBITION STN.
NOTTING HILL GATE
HOLLAND PARK

QUEEN'S RD.
KENSINGTON

NOTTING HILL GATE

HYDE PARK CORNER

...ERD'S ...USH
UXBRIDGE RD.

GARDENS
KENSINGTON PALACE

KNIGHTSBRIDGE

HIGH ST KENSINGTON

GLO'STER ROAD
SOUTH KENSINGTON

ALBERT HALL
VICTORIA ... MUSEUM

BELG...

...MITH ...MPIA

ADDISON ROAD

BROMPTON ROAD
VI...

EARLS COURT

SLOA... SQUA...

WEST BROMPTON

CHELSEA

CHELSEA FOOTBALL GROUND

FULHAM ROAD

THAME...

WALHAM GREEN

BATTERSEA PARK

PARSON'S GREEN

RIVER

3 D
"HOW TO GET THERE"
BISHOPSGATE TO KNIGHTSBR...

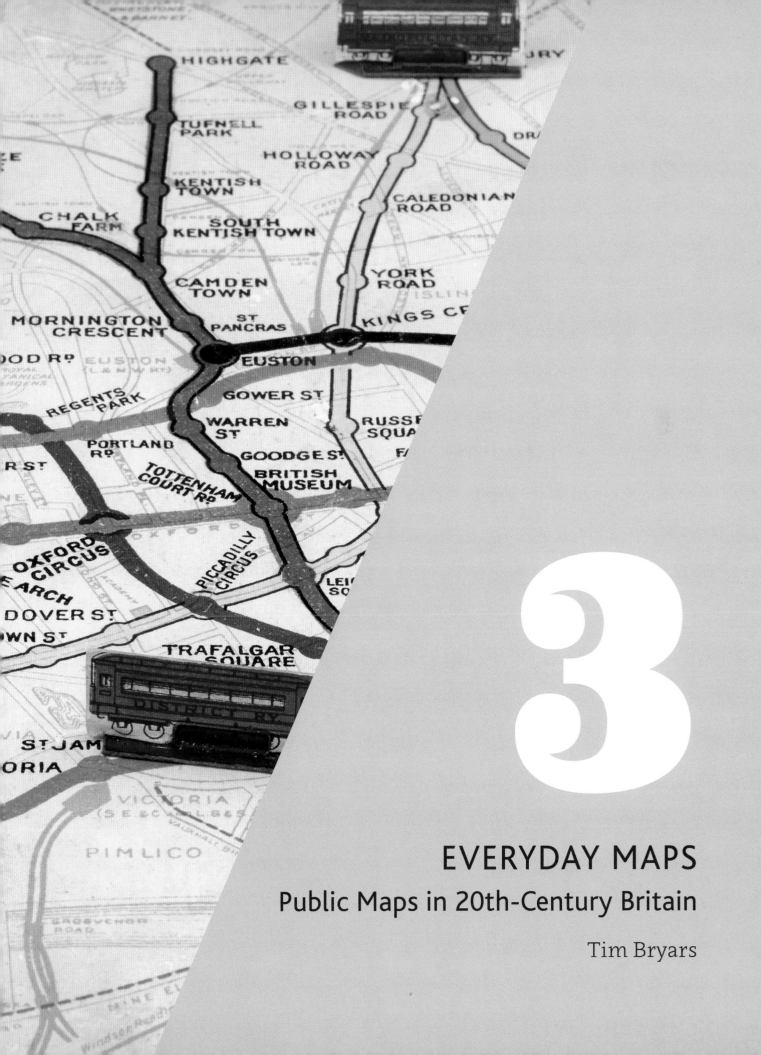

3

EVERYDAY MAPS

Public Maps in 20th-Century Britain

Tim Bryars

BRITISH RAILWAYS
PROPOSED WITHDRAWAL OF PASSENGER TRAIN SERVICES

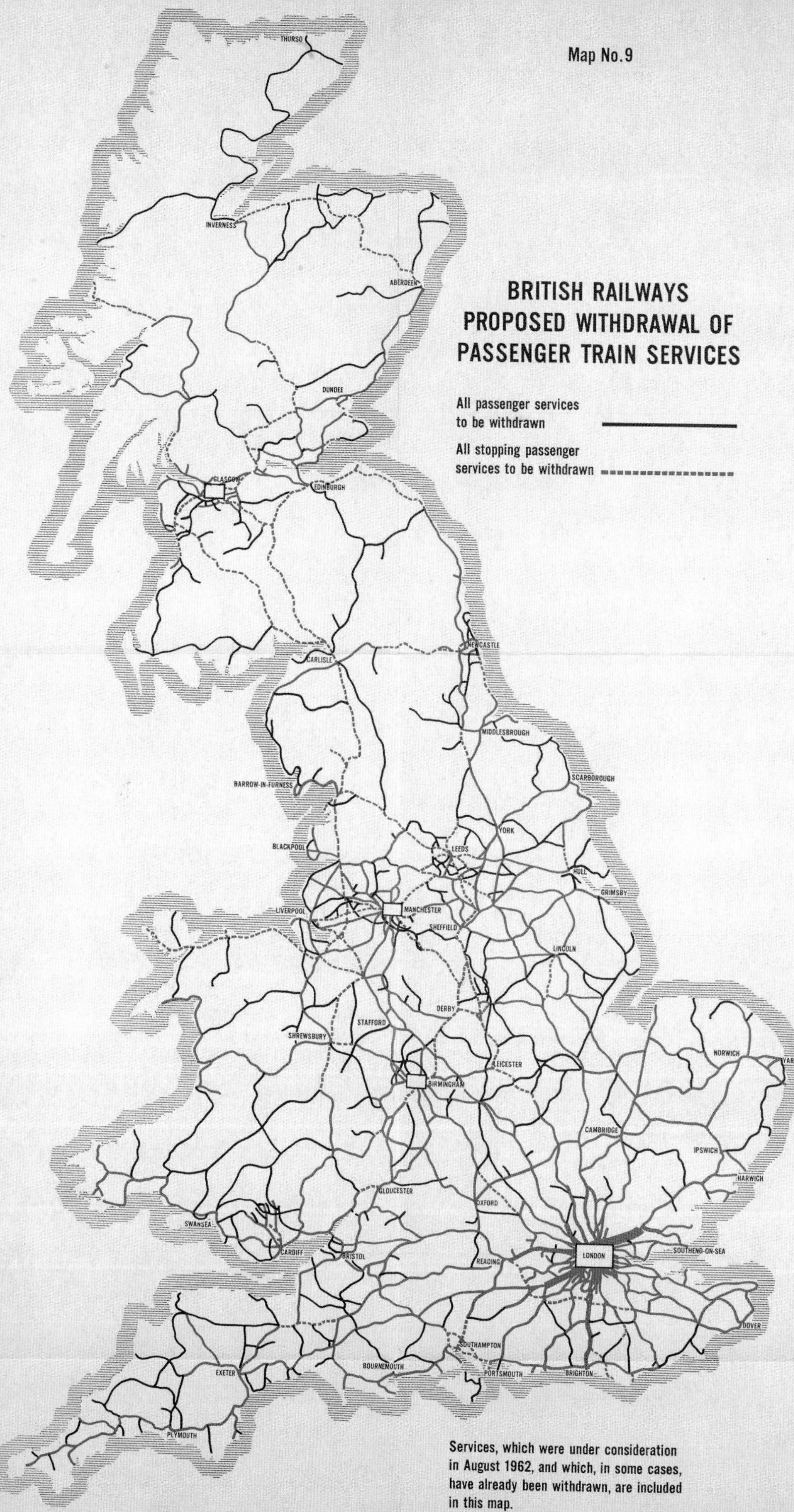

All passenger services
to be withdrawn ————————

All stopping passenger
services to be withdrawn ---------------------

THURSO

INVERNESS

ABERDEEN

DUNDEE

GLASGOW   EDINBURGH

CARLISLE   NEWCASTLE

MIDDLESBROUGH

SCARBOROUGH

BARROW-IN-FURNESS

BLACKPOOL   YORK

LEEDS   HULL

GRIMSBY

LIVERPOOL   MANCHESTER

SHEFFIELD

LINCOLN

DERBY

STAFFORD

SHREWSBURY   LEICESTER   NORWICH   YARMOUTH

BIRMINGHAM

CAMBRIDGE   IPSWICH

HARWICH

GLOUCESTER

OXFORD

SWANSEA   READING   LONDON   SOUTHEND-ON-SEA

CARDIFF   BRISTOL

DOVER

SOUTHAMPTON

EXETER   BOURNEMOUTH   PORTSMOUTH   BRIGHTON

PLYMOUTH

Services, which were under consideration
in August 1962, and which, in some cases,
have already been withdrawn, are included
in this map.

Now when I was a little chap I had a passion for maps. I would look for hours at South America, or Africa, or Australia, and lose myself in all the glories of exploration. At that time there were many blank spaces on the earth, and when I saw one that looked particularly inviting on a map (but they all look like that) I would put my finger on it and say, 'When I grow up I will go there.'

Joseph Conrad, *Heart of Darkness*, **1902**

A chance encounter with a different map, displayed in a Fleet Street shop window, seduces Conrad's protagonist Marlow – leading him inexorably towards deranged ivory trader Kurtz, two months' journey up the Congo River. The map of Africa 'had ceased to be a blank space of delightful mystery' as he remembered it from childhood, 'but there was in it one river especially, a mighty big river, that you could see on the map, resembling an immense snake uncoiled … it fascinated me as a snake would a bird'.[1] Conrad captures several elements central to everyday encounters with maps in the twentieth century, including assumptions about the continual 'improvement' and accuracy of maps: the filling in of blank spaces which were anything but blank to the people living there already. The most striking feature of this passage, though, is the ease of Conrad's assumption that everyone would be familiar with maps, even children: it shows just how widespread an awareness of maps had become, and illustrates their power to fire the imagination.

Conrad's text, written at the turn of the twentieth century, describes a reaction to maps that would have echoed with most people a century later. Maps were adapted to reflect new concerns and controversies, but people continued to make maps showing all or part of the Earth's surface for very much the same reasons as they had hundreds of years earlier. What changed in the twentieth century was not so much the depth of mapping as the breadth. Advances in printing and production, more so even than advances in surveying and mapping techniques, coincided with mass education to ensure that by the 1900s the stage was set for maps to reach into every part of daily life. As in previous centuries, the principal catalysts were warfare, improved living standards, travel and leisure opportunities; however, in the twentieth century, maps adapted to mirror some major changes that were taking place. After 1900, conscription (and aerial bombardment) meant that warfare was no longer confined to a small number of professionals; and, while most travel was still undertaken either for work (the first use of the word 'commuter' by a non-American writer, according to the *OED*, was by Rudyard Kipling in 1902) or (as in migration) for survival, it was only in the twentieth century that travel for other

59
Map no. 9 from *The Reshaping of British Railways*, also known as the Beeching Report. HMSO, 1963.
B.S. 122/58

60
Estra Clark, *A Map of Yorkshire*
Produced by British Railways, 1949.
Maps CC.6.a.35

reasons ceased to be the exclusive preserve of the rich and became widespread across all classes. The twentieth century ushered in the atom bomb at more or less the same time as the package holiday and the family car.

Just as the whole population became directly affected by war and travel, so the public became more familiar with maps than they had ever been before. Not just content evolved, but style as well: from journalistic cartography to advertising, the public was bombarded with and came to appreciate both conventional, surveyed maps, and map imagery, the latter heavily influenced in its turn by changing fashions in graphic design. Widespread assumptions about the neutrality and scientific accuracy of maps, reinforced in the schoolroom and by popular culture, rendered them ideal tools of persuasion: the appetite for maps was not driven exclusively by a desire for greater accuracy, and the majority of twentieth-century maps were not created for the purpose of navigation. Target audiences could be relied upon to recognise basic cartographic symbolism, born out of familiarity with maps as everyday household objects. Paradoxically, however, the very universality of maps – now appearing on brochures and posters, postcards and souvenir handkerchiefs, and commonly used to establish background in films and fiction – began to cloak them with invisibility: they had become part of the mind's furniture.

In terms of everyday map use for the majority of the population, the twenty-first-century phenomenon of the digital revolution is just as significant as the invention of printing. However, it is easy to forget that the real changes had already taken place. The huge increase in cheap printing, and later photocopying, from Conrad's time to the millennium meant that it had never been easier for individuals and special-interest groups to create and use their own maps, decades before digital technology was affordable and widely available. The twentieth century was indeed the heyday of the printed map.

More maps were printed in the twentieth century than in any other and – given recent advances in digital technology – that record is unlikely ever to be broken. Lamenting the passing of the paper map would be premature, however: sales of paper Ordnance Survey maps rose in 2014, and in 2015 Britain's mapping agency purchased a 25 per cent shareholding in Dennis Maps, 'one of the country's most significant large format litho-printers'.[2] However, for sheer volume and variety the twentieth century deserves to be celebrated as the pinnacle of a process which began with the earliest printed Ptolemaic maps in the 1470s. The first printed map of London was a gift fit for a monarch; 400 years later maps commemorating royal events were given away to schoolchildren. The twentieth century was the first age of near-universal map literacy, and for the first time maps were used and understood for almost every imaginable purpose by people of most social and educational backgrounds. Beginning in the nineteenth century, mass education coincided with the introduction of cheaper printing techniques, allowing for the mass production and distribution of material – some of it not merely cheap but free. As the twentieth century progressed, new technologies, first aerial photography then satellite imagery, allowed for the accurate and economical mapping of vast swathes of the planet, some of which had never been mapped before; it was a century when, partly through colonial and post-colonial mapping projects, the norms of Western mapping gained worldwide dominance.[3]

Maps and atlases had become increasingly affordable over the course of the nineteenth century. For 350 years the process of reproducing a high-quality image remained more or less unchanged: etching or engraving in reverse onto a metal printing plate, usually copper, from which black-and-white images could be pulled from a hand press, one at a time, on to hand-made paper, and to which colour could be added, by hand, if required. The investment in materials and skilled labour was considerable, and in the hand-press era maps were prestigious items. From the second decade of the nineteenth century onwards, technological advances in paper manufacture, mechanised printing, edition binding and the development of new reprographic techniques combined to reduce costs. Examples of relatively prestigious atlases, such as G. W. Bacon's *Large-Scale Atlas of London & Suburbs*, which sold for 25 shillings in 1913,[4] were hand coloured (albeit using stencils) right up until the First World War, but well before the turn of the century colour-printed lithography had contributed to the triumph of the cheap book movement, and basic sixpenny atlases from major publishers such as W. & A. K. Johnston, Cassell or Warne had become a staple of the schoolroom. Interpreting the relative values of historical commercial transactions is complicated, but the evolution of retail prices does support this downward trend.[5] In the twentieth century the price of 'war maps' for the domestic market – popular with Victorian audiences tracking Queen Victoria's 'small wars' in far-flung corners of the globe – rarely exceeded one shilling in their most basic paper form. Many were published in association with major newspapers. For example, the *Daily Mail* published George Philip & Son Ltd's map of the Boer Republics, 'to illustrate the present crisis in the Transvaal', in over twenty editions from 1899 to 1900, priced at one shilling. In 1914 a *General War Map of Europe* by the same maker, flanked by graphics showing the rival strengths of the Great Powers, was even more modestly priced at sixpence. The *John Bull War Map & Guide*, published by Odhams Press in 1914, was sold for one penny.

More luxurious and specialist cartographic publications continued to be produced. For example, the magnificent national atlas of Britain published by Oxford University Press in 1963 cost the enormous sum of twenty-five guineas[6] or thirty times the price of a typical hardback book; the latest James Bond novel, *On Her Majesty's Secret Service*, was published at sixteen shillings by Jonathan Cape in the same year. However, the twentieth century was the first in which maps, for so long the preserve of a wealthy elite, began to be given away free – something which, in the digital age, we take for granted at the point of use. From 1902, for example, the Central London Railway (the future Central Line, then an independent company) gave away 'a new form of map' to passengers, printed on paper and folding into a convenient format.[7] In 1906 the newly formed Underground Electric Railways Company of London (UERL) followed the same model. Print runs could be substantial, too – the first edition of Harry Beck's famous diagrammatic map of the London Underground, issued free in January 1933, had a run of 750,000 copies.

Maps and atlases were still perceived as having value, which made them suitable promotional gifts to commemorate special occasions. For example, *The Royal Primrose Atlas of the British Empire* was given away by royal warrant holders John Knight Ltd of the Royal Primrose Soap Works, to mark George V's Silver Jubilee in 1935. There was an obvious

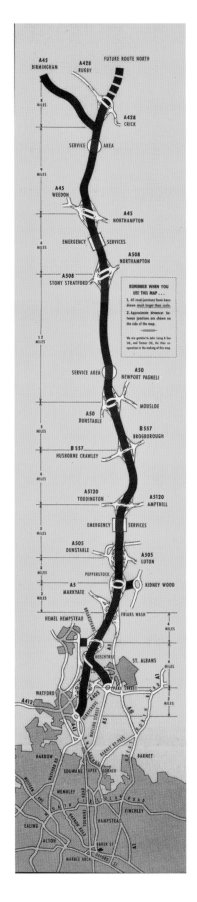

association between the Shell Oil Company and a linear map of the newly opened first section of Britain's first motorway, the M1, issued in 1959 (ill. 61). The habit of giving away promotional maps became deeply ingrained, and occasionally made for strange bedfellows. In the 1970s Duphar Laboratories Ltd issued a series of Bartholomew's regional maps with customised covers, tastefully devised in shades of brown, to promote Colofac (which according to the text on the map 'acts directly to relieve the symptoms of intestinal spasm, colicky pain and disturbed bowel function'); these were given away by sales reps as an inducement to doctors and other medical staff.

Cheap maps, large print runs and the use of maps in advertising presuppose a wide audience, and it is worth pausing to consider the extent and nature of map literacy. It has been suggested that 'from the 1920s it became a commonplace of children's literature to show children confidently using and making maps, and in doing so it was perfectly aligned with developments in geographical education and open air culture in general'.[8] Maps had featured in children's books, sometimes for the very young, even earlier: A. A. Milne's Hundred Acre Wood in the Winnie The Pooh stories[9] was presaged by Rudyard Kipling's 'inciting map of the Turbid Amazon'[10] which 'hasn't anything to do with the story' – making it all the more imperative that it would be understood and enjoyed by its young audience.

Showing actual map use came later, the outstanding example of the genre being Henry Deverson and Ronald Lampitt's *The Map that Came to Life* (ill. 62), published by Oxford University Press in 1948 at the price of eight shillings and sixpence, and reprinted and revised over the next twenty years. Lampitt was a commercial artist noted for his rural landscapes and scenic views of towns, some of which featured on railway posters.[11] He collaborated again with his brother-in-law, *Picture Post* picture editor Harry Deverson, on other juvenile literature, and he reworked the themes of *The Map that Came to Life* for the Ladybird series (*Understanding Maps*, 1967) – but only the large format and superb colour lithography of the OUP publication did full justice to his artwork. Two children, John and Joanna, set out on a walk with their dog Rover from their Uncle George's farm, using an Ordnance Survey map to guide them. On each page an enlarged section of the map is set alongside a bird's-eye view of the landscape it represents. The relationship between the two is cleverly managed in the text. On one spread the children decide 'it would be very interesting to lay the map out at the top of the hill and compare it with the actual scene'. Features on the map are realised in the image: the hill itself; the railway with its tunnel, viaduct and embankment; the water mill and its weir and part of the local reservoir. The locomotive steaming out of the tunnel, the ducks on the weir and the kestrel hovering above the reservoir are among the incidental details which bring the map to life.

It is tempting to view *The Map that Came to Life,* aimed at nine- to twelve-year-olds,[12] as a book for and about middle-class children, rather than as a book which taught a generation to read maps. There were cheaper alternatives, such as Puffin Picture Book 67, *About Maps* by Peter Hood, which cost two shillings in 1950. However, the middle-class antecedents of the central characters are unsurprising: working-class children were seldom the heroes of children's books before the 1970s.

61

Shell Oil linear map of the first section of the M1 motorway, 1959. Private collection

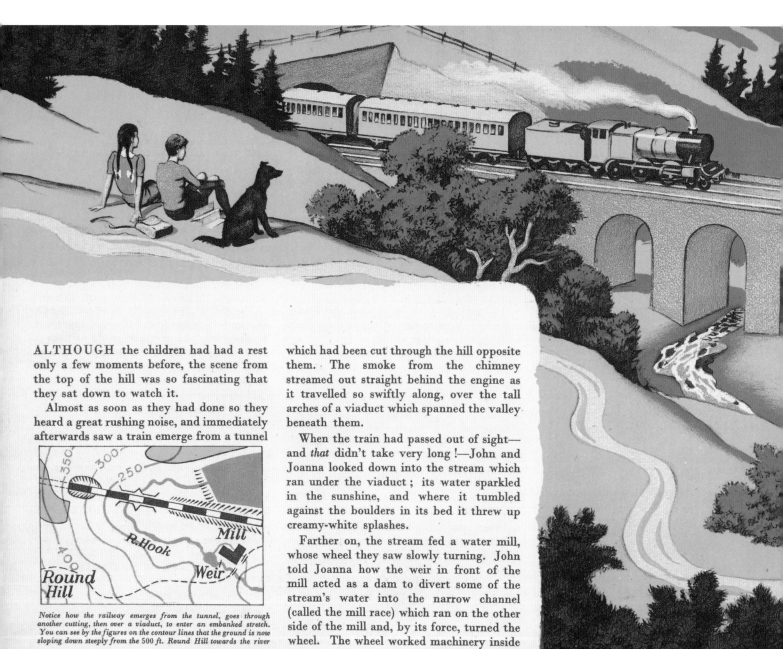

ALTHOUGH the children had had a rest only a few moments before, the scene from the top of the hill was so fascinating that they sat down to watch it.

Almost as soon as they had done so they heard a great rushing noise, and immediately afterwards saw a train emerge from a tunnel

Notice how the railway emerges from the tunnel, goes through another cutting, then over a viaduct, to enter an embanked stretch. You can see by the figures on the contour lines that the ground is now sloping down steeply from the 500 ft. Round Hill towards the river

which had been cut through the hill opposite them. The smoke from the chimney streamed out straight behind the engine as it travelled so swiftly along, over the tall arches of a viaduct which spanned the valley beneath them.

When the train had passed out of sight— and *that* didn't take very long !—John and Joanna looked down into the stream which ran under the viaduct ; its water sparkled in the sunshine, and where it tumbled against the boulders in its bed it threw up creamy-white splashes.

Farther on, the stream fed a water mill, whose wheel they saw slowly turning. John told Joanna how the weir in front of the mill acted as a dam to divert some of the stream's water into the narrow channel (called the mill race) which ran on the other side of the mill and, by its force, turned the wheel. The wheel worked machinery inside the mill which was used for many purposes, including grinding corn to make flour. In the mill pool the children could see the ducks floating.

There was a bridge to the mill and another which crossed the mill race to join the opposite bank. Beyond the mill was a meadow where some Shorthorns grazed; then the railway embankment; then some trees and, beyond these, a stretch of still blue water, with more trees beyond it. Joanna discovered by looking at the map that the stretch of blue water was a reservoir. (From this reservoir great pipes carried water all the way to Dumbleford and other more distant towns, for the townspeople to use when they turned their taps in their kitchens and bathrooms. Uncle George told them this afterwards.)

It *was* a fine sight, they agreed, and it was very interesting to lay out the map on the ground at the top of the hill and compare it with the actual scene.

Before they got up they saw a sparrow-hawk hovering high in the sky, searching for a small bird to swoop down upon. The children were very glad that it did not find one. (You can see the sparrow-hawk immediately above the water mill in the drawing.)

Then the children began to descend into the valley, passing sheep grazing on the sweet hillside grass which grew amongst the gorse bushes. Once, on their way down, a lark rose almost in front of their feet from its nest in a tuft of grass. It climbed higher and higher, singing its lovely song, until it vanished into the blue sky.

Soon the children were crossing the mill bridges to the sound of the waters of the mill race, and with Rover running on ahead they walked on by a grassy cart-track. . . .

62
Pages from *The Map that Came to Life*, by Henry Deverson and Ronald Lampitt. Oxford University Press, 1948.
Cup.1246.aa.53

It is difficult to establish the extent of map literacy across class boundaries. Charitable organisations such as the Society for the Diffusion of Useful Knowledge introduced maps to those without access to a formal education in the pre-Victorian era. In the first decades after the passing of the 1870 Elementary Education Act, compulsory school attendance for younger children had become more or less the norm across the country, and Britain had achieved near-universal basic literacy rates. Extracurricular activities and the teaching of languages, music and indeed anything other than basic literacy and numeracy were still frequently perceived as unnecessary or dangerously progressive, but 'investment in education for all was increasingly seen as a passport to prosperity rather than a threat to the nation's stability'.[13] History and a rudimentary geographical education had both become established as an accepted part of the curriculum in elementary schools by 1900. Laurie Lee recalled 'the old world map as dark as tea' which hung in his village schoolroom just after the First World War.[14] One late-Victorian pupil-teacher described how a map of England was used as a teaching aid: 'We used to stick a bit of cloth on it where it was Bradford, and a bit of carpet for carpet making areas and pin them on the map.'[15] There were few trained geography teachers and the subject remained a minority interest in secondary schools, with barely a toehold in universities.[16] However, respect for the subject was accelerated by the Boer War and the Great War, and after the passing of the 1902 Education Act the focus of interest shifted from elementary to secondary; there was undoubtedly a sense of optimism. The Geographical Association, with a meagre 121 members in 1900, launched its new journal, *The Geography Teacher*, on the first day of the new century in 1901; one of the first articles discussed using the Ordnance Survey in lessons.

Map reading could also be learnt outside the classroom. 'The Scout's badge is the arrow head, which shows the north on a map or on the compass. It is the badge of the scout in the Army, because he shows the way', wrote Baden-Powell.[17] *Scouting for Boys* also taught rudimentary pathfinding: 'practise map reading and finding the way by the map'.[18] Scouting and comparable organised youth movements such as the Boys' Brigade and Church Lads' Brigade were a response to the Edwardian debate on 'national efficiency', through which 'the leaders of these movements hoped to reverse the slow slide of Britain's position in relation to its continental rivals'.[19] Inculcating a sense of patriotism was central to the process, and in this context Baden-Powell considered that 'a map of the Empire is very desirable'.[20] One of the recommended maps was *The Navy League Map*, intended to demonstrate the centrality of British sea power to the future of Britain and its empire (ill. 63).[21] A handbook which accompanied the map was dedicated by its schoolmaster author 'to the British schoolboy'[22] and a preface by the Earl of Meath proposed that this 'little work should be in the hands of every young Briton, and no school should be considered properly equipped which has not the full-sized Navy League Wall Map of the Empire hanging on the walls within easy view of the scholars'.[23]

Although in theory Scouting was open to all, factors such as methods of recruitment and the cost of regalia severely limited working-class participation. The harnessing of a generation of youth to the ideals of empire may only have been true of 'the middle and lower middle class boys who flocked to join Baden-Powell's Scouts, and even in this case, a simple cause and effect relationship between participation and indoctrination is difficult to prove'.[24] Nevertheless, the millions who volunteered for Kitchener's New Armies between 1914 and 1916 (the introduction of conscription) are likely to have had a residual memory of the 'British pink' covering large areas of the map on the class-room or Scout hut wall. This remained true until at least the 1950s. 'Empire Day' was the occasion for a school half-day holiday between 1904 and 1958.

The conscript armies of the two world wars were the first in British history. The Boer War of 1899–1902 saw civilians in uniform – there were a large number of volunteers among the 250,000 British and Commonwealth troops who took part, and there was a correspondingly large appetite for souvenir maps at home, such as Stanford's pair of maps charting the relief of Ladysmith, printed some months after the event.[25] The service men and women who volunteered or were called up in 1914–19 and in 1939–45 (and indeed, as National Servicemen, until 1963) represent a full cross-section of British society; the impact on map-mindedness, a widespread recognition of the value of maps, was correspondingly immense.

During the Great War, manuals of map reading were prepared 'for the instruction of officers and non-commissioned officers',[26] though their numbers were greatly swollen by wartime expansion, and in practice maps were occasionally issued 'to every man in a sub unit'.[27] Lectures and other educational activities to keep men occupied when they were out of the trenches were mostly organised under the auspices of the Young Men's Christian Association and similar bodies until the very end of the war, when official interest was roused by the prospect of mass demobilisation.[28] However, soldiers were perhaps as likely to encounter maps behind the line as in it. As part of his war work E. W. Hornung, creator of Raffles – the 'amateur cracksman' or gentleman-thief – describes setting up a YMCA library in Amiens, just behind the front (it was overrun within months, during the German Spring Offensive of 1918). Troops could wipe the mud of war from their feet, but 'the first thing to be seen inside ... was the complete *Daily Mail* sketch-map of the Western Front, the different sheets joined together and ... the Line ... pegged out from top to bottom with the best red-tape procurable in town. It toned delightfully with the art-green of the sketch-map.'[29] As with school and Scouting, the extent of participation is difficult to establish: 'it was the better-educated men who came flocking in, the intellectual pick of an Army Corps who made our hut their club.'[30]

During the Second World War education for all ranks played a greater official role. Map-reading exercises are a staple of military memoirs and semi-fictionalised accounts of basic training and life in barracks. George MacDonald Fraser's short stories about life in a Highland Regiment in early post-war North Africa include 'The Constipation of O'Brien': his platoon was dropped from a lorry in pairs, at night, with nothing but a map, a compass and matches, with the objective of evading (or overpowering) another unit, defending a bridge marked with a red lamp.[31] Spike Milligan recalls a similar exercise in southern England. Ordered to set up an observation post 'at Map Reference 8975-4564' Milligan's platoon found itself at the bottom of a 'deserted chalk quarry'. His platoon commander 'consulted his map. "There must be something wrong," he said, looking intelligent at two hundred feet below sea level. According to my calculations we should be on

63 (next page)
The Navy League Map: Illustrating British Naval History
W. & A.K. Johnston, 1901.
Maps 950.(138)

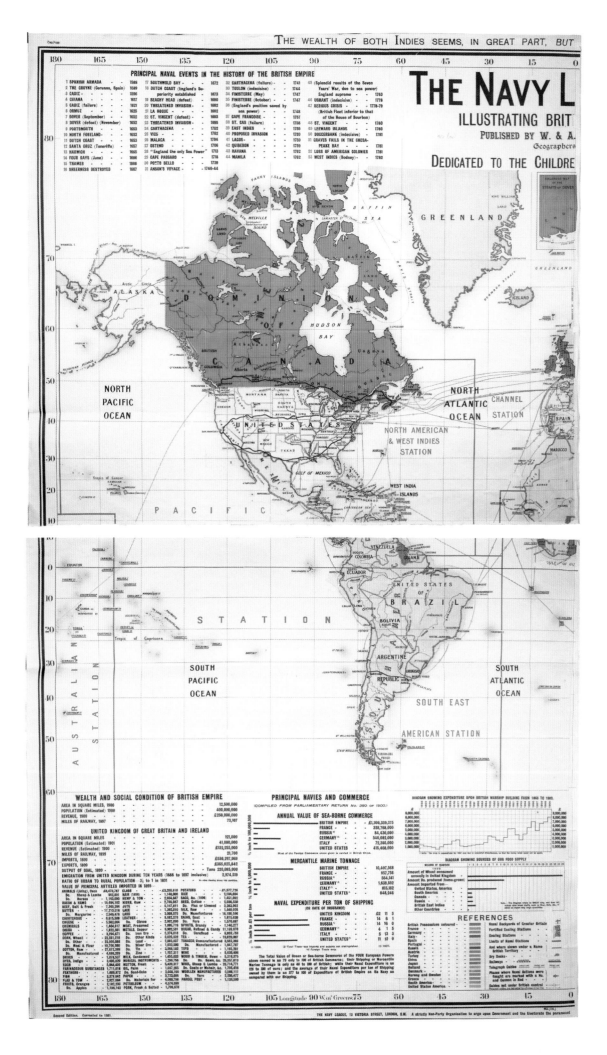

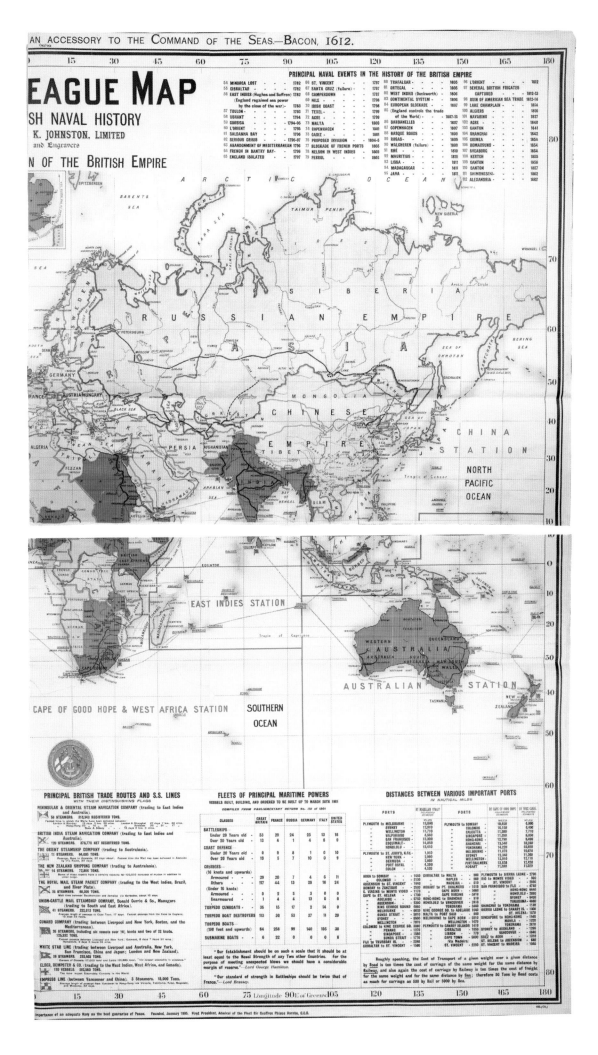

top of a hill, looking down a valley.' Gunner Milligan said, 'But we aren't on top of a hill looking down a valley, are we sir?' 'No we're not Milligan. How shrewd of you to notice. This could mean promotion for you, or death.'[32]

Adult education under the auspices of the Army Education Corps (AEC) was more structured during the Second World War than it had been in the First. To an extent this may have been prompted by the sheer numbers of troops who needed to be kept occupied (for every combat soldier there were, on average, eight in support).[33] For this generation of civilians in uniform, as with the population at large, there was also a sense that the empty promises of 'homes fit for heroes' made during the First World War would not be repeated: victory would be followed by real change for the better. In 1941 the Army Bureau of Current Affairs (ABCA) was established to inform service men and women of all ranks how the war – presented as a defence of democracy – was progressing, and to encourage debate on how to build a better democracy after the defeat of fascism. Regimental officers were supplied with alternating fortnightly bulletins, *War* and *Current Affairs*, and on the basis of these instructed to give talks and lead discussion groups with their men, on topics such as 'the British way and purpose', at a platoon or company level.[34] Maps were prominent among ad hoc visual aids prepared by trained AEC staff: 'outline maps in two colours with no complicated physical features or elaborate details to mislead the readers ... the day's news taped to an outline map of the battle front with coloured ribbons'.[35] Another 'popular feature'[36] which survived in service life for some time after the war was the *ABCA Map Review*. Edited by Lionel Birch, a former assistant editor of *Picture Post*, and also issued fortnightly, this was a large double-sided poster which employed simple maps and supporting photographic images to set the major events on all fronts in a geographical context.[37]

The world wars of the twentieth century mobilised nations, not just armies. Some military maps were simply repurposed for civilian life. The 1940 and 1941 'War Revision' editions of the Ordnance Survey, though printed using the War Office Cassini Grid for reference, were perfectly suited to normal civilian use, and indeed from early 1943 were actually sold to the public (with 'sales copy' stickers, presumably to deter the unofficial resale of military stock). Some reuse of military mapping was more creative. Purchased as government surplus after the war or given as gifts by friends in the services, fabric 'escape and evasion maps' were sometimes transformed into underwear. A rayon camisole made from recycled maps of the Far Eastern theatre has found its way into the Museum of London, and the Imperial War Museum has preserved a set of underwear made from a silk map of Italy for the elder daughter of Earl Mountbatten of Burma (ill. 64).[38]

More than any amount of official education or redistribution of military mapping, it was the global scale of twentieth-century conflicts which generated a surge in map use. For millions of Britons, war brought the first experience of foreign travel. Previously unheard-of French and Belgian towns assumed a permanent place in the national consciousness. More distant places – some of them, like Jerusalem or Damascus, names dimly recalled from Sunday school – assumed a new reality. Individual units created souvenir maps and even Christmas cards tracking their own, very personal, progress across the theatres of war (ill. 65). For operational reasons, service personnel

64
Underwear made from silk escape maps, owned by Countess Mountbatten, 1940s.
Imperial War Museum, EPH 10930

MAPS AND THE 20TH CENTURY: DRAWING THE LINE

Wishing you a
Merry Christmas and a
Happy New Year.
TO
My Nephew Willie
FROM
Your Loving Aunt Lizzie
1915 to 1916

65
ANZAC Christmas Card, 1915–16.
H.R.J.Series.
Maps CC.5.b.51

We can send our Christmas greeting
With pride as never before,
For who can forget that landing
On Gallipoli's treacherous shore?
When the Sons of the South bore witness
That they would fight as never before,
And with might and main would still maintain
England's cause in her righteous war.

H.R.J. Series—British Production.

MAP OF THE BOMBING OF CANTERBURY

REPRINTED IN AID OF

R.A.F. PILOTS AND CREWS FUND

(Registered under War Charities Act, 1940)

CITY AND COUNTY OF CANTERBURY

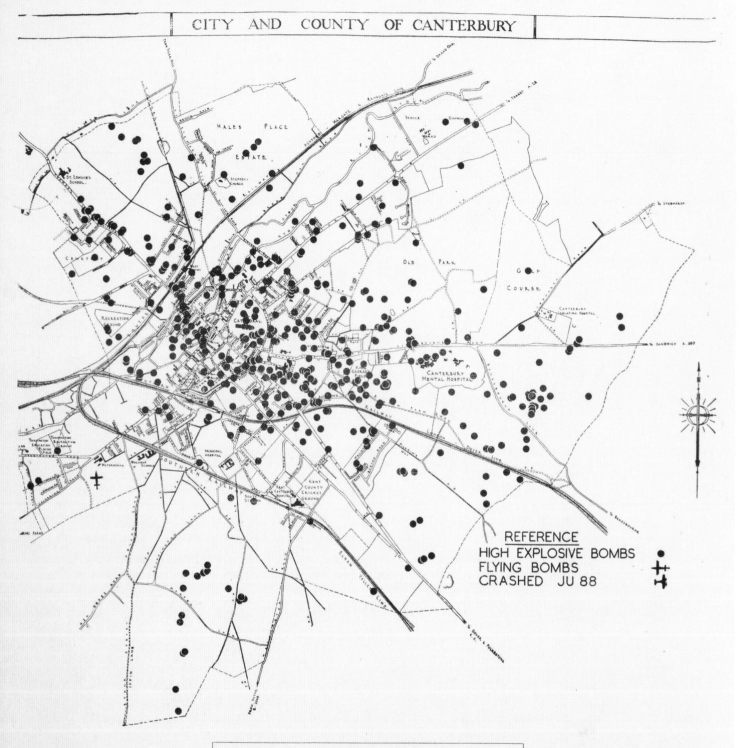

REFERENCE
HIGH EXPLOSIVE BOMBS
FLYING BOMBS
CRASHED JU 88

PRICE 6d.

PRICE 6d.

THE FOLLOWING DETAILS OF THE RAIDS ARE COMPILED TO 30th SEPTEMBER, 1944

ALERTS	2,477
RAIDS	35
H.E. BOMBS	445
INCENDIARIES	10,000
FATAL CASUALTIES	115
SERIOUSLY INJURED	140
SLIGHTLY INJURED	240
PROPERTIES DEMOLISHED	808
PROPERTIES DAMAGED	6,738

In the June 1, 1942, Baedeker raid, there were 182 H.E.'s and nearly 8,000 incendiaries dropped, and the casualties were 50 killed, 55 seriously injured and 74 slightly injured.

On Saturday afternoon, October 31, 1942, during the daylight raid by low-flying F.W. 190's, 29 H.E.'s were dropped, 33 people were killed, 49 seriously injured and 61 slightly injured. That night 23 H.E.'s fell in a further raid, together with over 500 incendiaries, killing two people, seriously injuring five and slightly injuring one.

Printed at the Office of "The Kentish Gazette & Canterbury Press."

were prohibited from passing on detailed information about their whereabouts, but efforts were sometimes made to evade the military censors so that family and loved ones could place them on the map. Perhaps not all of them were as elaborate as that employed by one of Angela Brazil's schoolgirl heroines, Winona Woodward, who deciphers letters from the front sent by her brother Percy using a prearranged cryptogram: 'she would find the spot on the big war-map that hung in the dining-room and mark it with a miniature flag, feeling in closer touch with him now she knew exactly where he was located.'[39] Swept up by world events, cartography afforded a degree of connection between those at the front and those left behind.

At the same time, maps were also produced in an attempt to make sense of the dangers of the home front. At the end of the Great War Harmsworth printed maps showing air raids and naval bombardments across Britain, alongside maps of the other theatres of war, which formed a supplement to *Harmsworth's New Atlas*.[40] During the Second World War some local newspapers printed bomb damage maps, especially after the launch of V-1 rockets against London, many of which came down over south-eastern England during the second blitz of 1944. A map of Canterbury published by the *Kentish Gazette* marks the location of 445 high-explosive bombs, the flying bomb which landed near the waterworks, and the Junkers Ju 88 bomber which crashed near the cathedral (ill. 66). One hundred and fifteen people had been killed, almost half of them in the so-called Baedeker raid of 1 June 1942, and the map is an expression of civilians' new place on the front line in the age of aerial warfare.[41]

During the Great War there was a sharp increase in the sale and circulation of newspapers, which 'more than any preceding event, demanded the use of maps on a hitherto unimagined scale ... almost certainly connected to the strict censorship of photography imposed, with the tacit approval of the press, for the duration of the conflict'.[42] Photographic images proliferated – maps represented 90 per cent of the visual images in *The Times* before 1919, and perhaps 10 per cent in the interwar period – but this may have been because they were for the first time readily available, and easily reproduced in newsprint, rather than because they had superseded the map.[43] Traditional bird's-eye or balloon views were influenced by stylised perspectives derived from the reality of air travel. The 'infographic' is nothing new: it would have been as familiar to an Edwardian scanning the *Illustrated London News* or *The Graphic* for information, say, about the Japanese siege of Port Arthur in 1904 as to a reader of the British press in the 1990s, during the invasion of Iraq. Maps were frequently combined with insets showing key people and places, and even examples of military hardware. However, the overall range and quality of newspaper maps improved dramatically during the Great War, and by the 1920s many newspapers either maintained their own cartographic units or had established close relationships with commercial firms offering the same service. Journalistic cartography was to remain an influential feature of twentieth-century life.

This journalistic cartography bleeds into other forms of politically persuasive mapping. Over the course of the twentieth century the UK achieved universal suffrage (the franchise was extended in 1918, 1928 and 1969), and for all parliamentary elections commercially published maps enabled readers to record the results as they became

66

Map of the Bombing of Canterbury, reprinted in aid of the R.A.F. Pilots and Crews Fund. Printed at the Office of the Kentish Gazette and Canterbury Press, 1945. Private collection

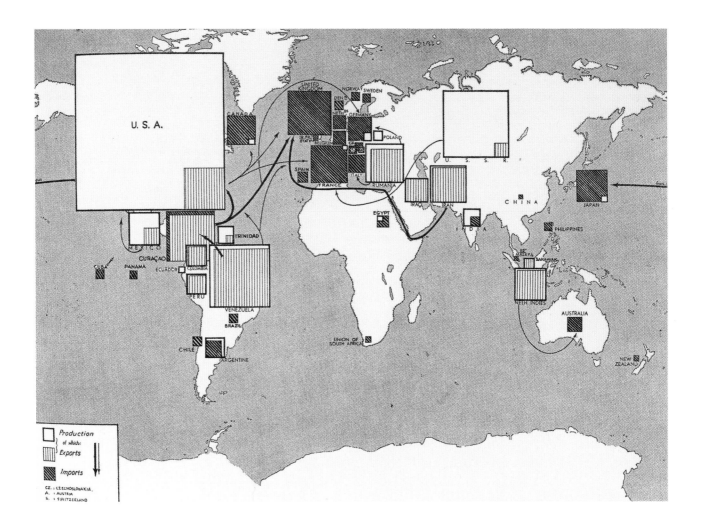

67
Alexander Radó, 'Petroleum', from
The Atlas of To-day and To-morrow.
Victor Gollancz, 1938.
Maps C.44.d.96

available. Sometimes, as with *The Times* election map for 1964, it was a simple matter of colouring in.[44] For the 1906 election, which resulted in a Liberal landslide, the *Illustrated London News* was more creative: 'in the margin we give a race-course and two jumpers, a Unionist and a Radical. These may be detached and pinned on the wall, and the figures shifted point by point according to the gains of either party.'[45]

Much journalistic cartography was more openly slanted. *The Atlas of To-day and To-morrow* – presented as a 'cartographic Inventory' of the 'political, economic and social problems' of the day –[46] is the major work in English by the Hungarian Jewish Soviet intelligence agent and cartographer Alexander Radó (ill. 67). It included 'strikingly innovative maps with a broadly socialist message prepared by Martha Rajchman, a Polish Jewish artist who had trained in Paris ... The atlas sold more than 5300 copies'.[47] In Geneva, Radó had established the Geographical Press Agency, or Geopress, which prepared hundreds of detailed maps during the Spanish Civil War. These were then sold to newspapers and magazines around the world, demonstrating the 'rising demand for journalistic cartography as the international crisis worsened'. Geopress also demonstrates perfectly the crossover between academic cartography, news and information-gathering and dissemination, and intelligence: 'Geopress provided the perfect cover for Radó's intelligence activities.'[48]

Radó's atlas was published by Victor Gollancz, an inspired judge of commercial fiction and founder of the Left Book Club, which was to become 'something of a legend in British life', sometimes credited with having a 'significant influence' on the Labour Party's election victory in 1945.[49] The Left Book Club had 50,000 members by the outbreak of the Second World War, and from 1934 onwards an *Atlas of Current Affairs*, revised annually, had appeared in the Club's distinctive orange wrappers, marked 'not for sale to the public'.[50] The author (of text and maps) was the socialist writer and illustrator Frank Horrabin: 'no one can read a newspaper intelligently today without some background knowledge of world geography', he wrote, before 'urging' the reader 'to make his own marginal additions as and when his newspaper gives him additional information'.[51] Smaller left-leaning publishers continued the work of the Left Book Club, such as the Pluto Press, which published titles including *The State of the World* (1981) and *Women in the World: An International Atlas* (1986).

Horrabin's atlas covered contemporary controversies such as Jewish settlements in Palestine[52] using a simplified version of the map published by HMSO in the Hope-Simpson *Report on Immigration, Land Settlement and Development in Palestine* in 1930.[53] Throughout the twentieth century HMSO 'actively promoted access to government publishing', through the book trade, its own bookshops and mail order, reaching institutional customers including businesses and libraries, and the 'constantly shifting margin of individuals who have an interest in a publication, old or new'.[54] Official maps were by no means necessarily secret maps. Sales figures are difficult to establish[55] but some publications undoubtedly enjoyed a wide circulation. Patrick Abercrombie's proposals for the post-war reconstruction of London, the Beeching Report (with its notorious map 9, 'Proposed Withdrawal of Passenger Train Services') and its contemporaneous counterpoint, the Buchanan Report, which argued that British cities should be redesigned to accommodate the motor car, were handsomely produced and survive in significant numbers.[56] Though often filtered through newspapers and other popular publications, such as Ernő Goldfinger's Penguin paperback championing the Abercrombie Plan,[57] the influence of HMSO mapping on public life was considerable.

Official mapping and journalistic cartography were also fused in some of the output of the Isotype Institute. Like the maps printed by Gollancz, the visually memorable Isotype (International System of Typographic Picture Education) charts created by Otto Neurath and his colleagues aimed to disseminate a left-leaning approach to current affairs among a wider audience.[58] Series such as *The New Democracy* were published commercially, during and just after the war, but charts also appeared in HMSO publications and also as separate publications, for example displayed on the walls of the surgeries and clinics of the new National Health Service.

Neurath conceived Isotype as a form of hieroglyphics for the twentieth century. As a separate but related category, cartograms use distorted geographical outlines – often derived from statistics – to convey social information with the aim of influencing opinion, typified by those in Danny Dorling's *New Social Atlas of Britain*.[59] They succeeded – despite being distorted – because in the twentieth century the public was not only educated to become map-literate but was also, outside school, exposed to a profusion of other striking graphic images. The process did not begin abruptly in 1900, but in the twentieth century the public became more graphically sophisticated (and less prejudiced) than previous generations. Writing his 'visual autobiography' in the 1940s, Neurath referred to his own times as 'the century of the eye'.[60]

Aside from newspapers and journalism, the revolution in transport and leisure opportunities brought millions of Britons into frequent contact with maps. The twentieth century witnessed great advances in living standards, neatly summarised by Ernst Gombrich: 'To be sure, life is still hard for many people … but most people who work in factories and even most of the unemployed live better today than many medieval knights must have done in their castles.'[61] The nonchalant pipe-smoking rambler consulting his map on the front of the 1929 Ordnance Survey map of Dorking and Leith Hill encapsulates many popular perceptions of twentieth-century map use (ill. 68). This is an example of the cover art painted by Ellis Martin for the Popular Series, which was an integral part of the Ordnance Survey's commercial drive after the Great War. In the background are cyclists, a private motor car and a bus. Aimed squarely at the leisure market, in direct competition with the hiking, cycling and motoring maps issued by commercial mapmakers – most of which were ultimately derived from the Ordnance Survey itself – the Popular Series was an attempt to make the government agency self-supporting in the face of the post-war economic slump.

The first assumption we are invited to make is that maps had indeed become popular: affordable and comprehensible to a large part of the population. The second assumption, every bit as important, is that maps are in some sense neutral or scientific representations of the Earth's surface, rendered to the best of the cartographer's abilities to inform, rather than lead, the reader. This second assumption was reinforced by education and in popular culture, underpinning the power of maps as tools of persuasion. Distortions on the map, whether consciously or unconsciously produced, were often overlooked by the reader.

During the nineteenth century the railways democratised travel, ushering in a new era of the daily commute, for an increasingly urban population, and escapes to the seaside and the country. Cheaper travel and cheap printing led to an explosion in the publication of guidebooks and travel literature, frequently leading to the mapping of places which had no strategic or commercial significance outside the burgeoning leisure industry. Progressing from authorised days off, 'paid holidays slowly became part of the general wage-work bargain' – and a right for 40 per cent of the workforce by the time of the 1938 Holidays with Pay Act.[62]

Maps go to the very heart of rail travel, representing the movement of goods or passengers by the fastest, most direct route. Dedicated mapping of the extent of the rail network, by independent commercial publishers or the railways themselves, has been part of the landscape since the first age of railway mania (Bradshaw's *Railway Map of Great Britain* was first published in 1839). As well as being given away in booking halls or pasted to platform walls, maps were sometimes built into the very fabric of stations. Glazed ceramic tile maps survive at London Victoria and Manchester Victoria, and roughly a dozen Edwardian maps manufactured by Craven Dunnill & Co. Ltd for the North

68
Ordnance Survey "One-Inch" Map: Dorking & Leith Hill. Cover illustration by Ellis Martin, 1929. Private collection

ORDNANCE SURVEY "ONE-INCH" MAP

Dorking & Leith Hill

Mounted on Linen
Price Three Shillings

Published by the Ordnance Survey Office, Southampton

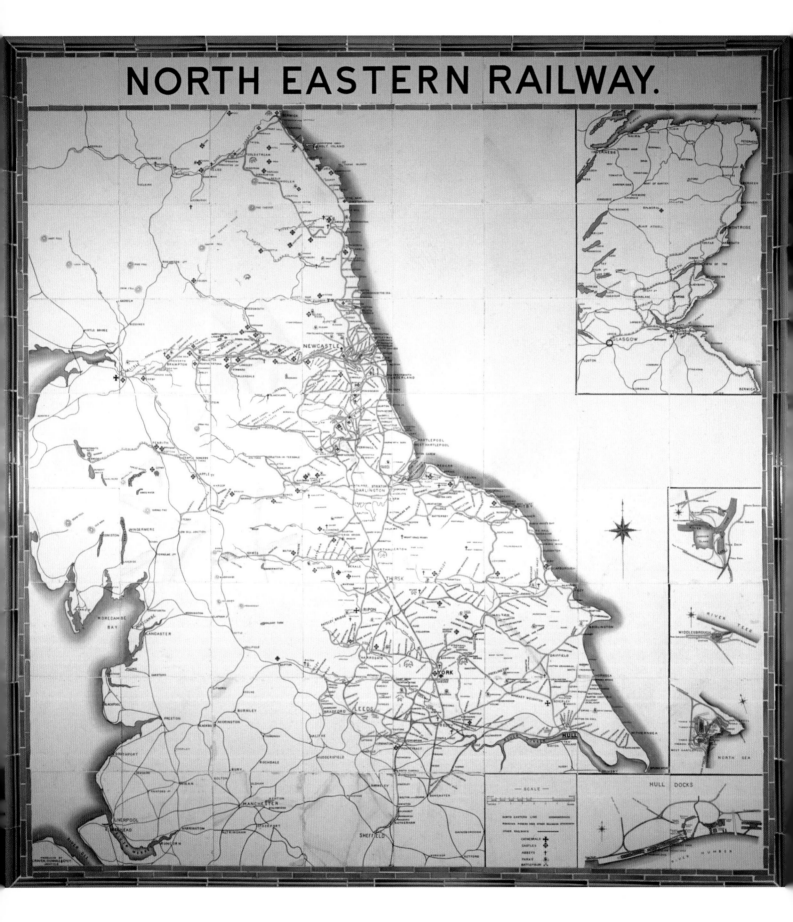

Eastern Railway are known to exist, nine of them in their original locations (ill. 69).[63] Practical in that they are durable and easy to clean (if not to update), maps like these embody the pride and confidence of the railway companies which commissioned them, as powerful even in their obsolescence as the Forma Urbis Romae which once dominated the Temple of Peace in ancient Rome.[64]

In the twentieth century the railways built on half a century or more of passenger mapping. The official LNER publication *On Either Side* was produced in the 1920s and 1930s, 'depicting and describing features of interest to be seen from the train' between London and Scotland. In style, the linear map is similar to maps by Henry Cole dating back to the early years of railway mania in the 1840s.[65] In content, it is not so far removed from the type of information provided to bored coach passengers in works such as Bowen and Owen's *Britannia Depicta* 200 years earlier. It is lavishly illustrated with images of locomotives and luxurious coach interiors, of cathedrals and castles, and the text is full of snippets of local history, feats of engineering and information about good trout fishing and other leisure activities. As such, this is an example of aspirational cartography: there is a strong implication that passengers on the LNER are the sort of people who enjoy luxury travel; similar motifs might be employed by Cunard or Concorde. The Mile-by-Mile series by Stuart Pike, covering routes on the 'big four' railway companies shortly before nationalisation, follows the same format though it was more cheaply produced, owing to the exigencies of post-war economy measures, and perhaps geared more towards the railway enthusiast in its technical detail (though a celebration of engineering feats was common in earlier maps). As well as route maps, the railways employed maps to draw in passengers making journeys for pleasure. MacDonald Gill's map of the *London Wonderground* (1913) inspired a generation of comic mapmaking. Kerry Lee's pictorial maps, for example, owing something to the style of Gill but with a vibrancy and inventiveness which is wholly distinctive, drew tourists to British towns between the late 1930s and the early 1960s (ill. 70). Lee created numerous designs for British Railways, and the British Travel Association distributed his maps, such as *London: The Bastion of Liberty*, through leading American stores, generating much needed foreign revenue and encouraging US tourists to visit Britain just after the war.

Mapmakers responded to new forms of transport as they emerged. Well before the turn of the century, companies as such G. W. Bacon & Co. Ltd, and George Philip & Son, were marketing maps aimed directly at cyclists, indicating potential hazards such as hills to be ridden down with caution (more flattering for the user than indicating hills too steep to cycle up) or where it was advisable to dismount. Civilian air travel become a feature of life for a privileged few (there were some 147,000 passenger journeys on British airlines in 1938, by which time private car and motorcycle ownership could be reckoned in the millions)[66] and led to the creation of some remarkable passenger maps; the 1934 Imperial Airways *Map of the European Air Routes* was illustrated by Edward Bawden, for example (ill. 71). Public transport methods in the form of buses, trams and trolleybuses all generated their own forms of mapping to meet passenger requirements. However, the twentieth century was predominantly the century of the motor car. While printed road atlases and maps in Britain were hardly new, stretching back to John Ogilby's

69
North Eastern Railway ceramic
tile map, 1903

70

Kerry Lee, *London Town*, 1951.

Private collection

LONDON TOWN

ISLINGTON

The Angel

BLOOMSBURY

Euston STATION

St. Pancras STATION

King's + STN.

Liverpool Street & Broad St. STATIONS

St. Paul's Cathedral

Holborn Viaduct STN.

Guildhall

THE CITY

The Bank

Charing + STATION

Blackfriars STATION

Cannon Street STN.

Fenchurch St. STN.

Waterloo STATION

The Tower of London

BOROUGH of SOUTHWARK

London Bridge STATION

St Katharine's DOCKS

Tower Bridge

Lambeth Palace

Trafalgar Square

LONDON – PARIS AIR ROUTE
(Croydon – Le Bourget .)
225 MILES
REFERENCE

Air Route	
Aerodromes	CROYDON ⊙
Prohibited Areas	
Frontiers	+++++++
Roads	
Railways	
Rivers	
Canals	

SCALE

10   0   10   20   30 MILES

KILOMETRES 10   0   10   20   30   40   50

*FIGURES IN MARGINS DENOTE DISTANCES IN MILES ALONG
ACTUAL LINE OF ROUTE.*

PRINTED IN GREAT BRITAIN &
PUBLISHED BY "RAYNOIL" MAPS LTD.
6, TAVISTOCK SQ. LONDON W.C.1

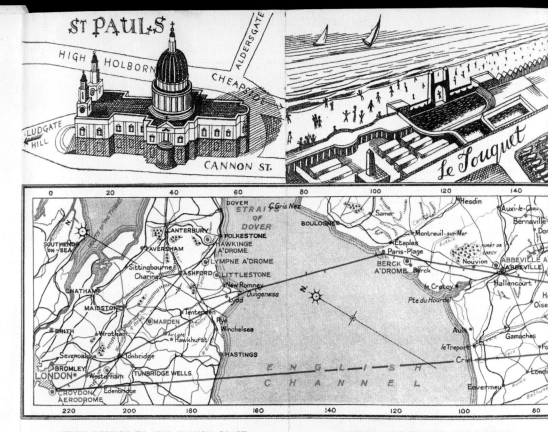

FROM LONDON TO THE FRENCH COAST

The route from Croydon crosses the wooded Kent Hills,
over hopfields and orchards and out over the cliffs on
the English Coast.
Here begins the Channel crossing to the French Coast,
where below may be seen ships ploughing their way up
and down the Straits of Dover and the English Channel

SEASIDE TOWNS AND FOREST LAND

At the French coast seaside resorts with their gay bathing
huts, fishing boats and dunes, catch the eye as the aeroplane
continues on the remaining 100 or so miles to Paris. The
countryside spreads itself out in a network of cultivated
fields and small woods interspersed with villages.
In general the countryside is of little interest on this sector

71
Edward Bawden, 'London – Paris
Air Route (Croydon – Le Bourget)',
Map of the European Air Routes,
Imperial Airways, 1934.
Private collection

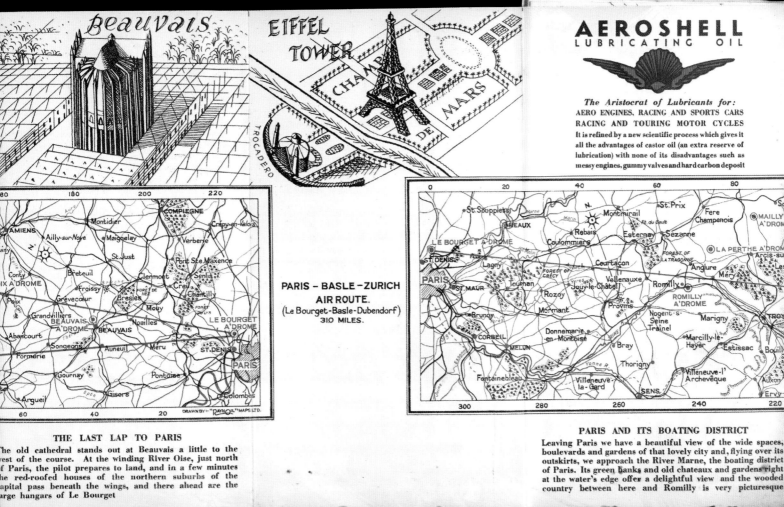

AEROSHELL
LUBRICATING OIL

The Aristocrat of Lubricants for:
AERO ENGINES, RACING AND SPORTS CARS
RACING AND TOURING MOTOR CYCLES
It is refined by a new scientific process which gives it
all the advantages of castor oil (an extra reserve of
lubrication) with none of its disadvantages such as
messy engines, gummy valves and hard carbon deposit

PARIS – BASLE – ZURICH
AIR ROUTE.
(Le Bourget - Basle - Dubendorf)
310 MILES.

THE LAST LAP TO PARIS

The old cathedral stands out at Beauvais a little to the
west of the course. At the winding River Oise, just north
of Paris, the pilot prepares to land, and in a few minutes
the red-roofed houses of the northern suburbs of the
capital pass beneath the wings, and there ahead are the
large hangars of Le Bourget

PARIS AND ITS BOATING DISTRICT

Leaving Paris we have a beautiful view of the wide spaces,
boulevards and gardens of that lovely city and, flying over its
outskirts, we approach the River Marne, the boating district
of Paris. Its green banks and old chateaux and gardens right
at the water's edge offer a delightful view and the wooded
country between here and Romilly is very picturesque

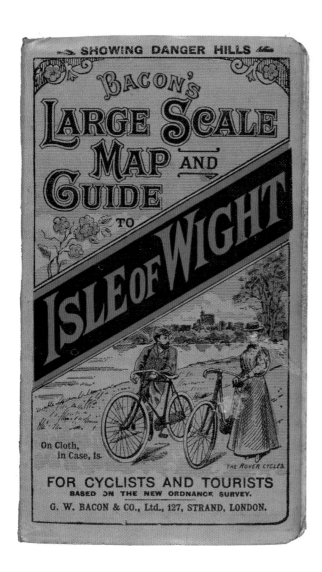

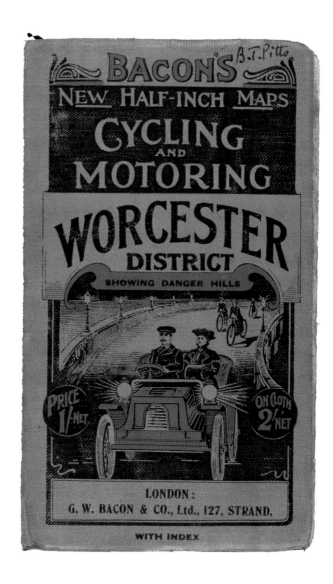

Britannia in the 1670s, transport had been speeding up ever since. The cycling maps of the late nineteenth century were rapidly adapted to encompass both cycling and motoring, and the oil companies were swiftly on the scene. Pratt's was among the first in 1904, with a motoring atlas aimed at chauffeurs and wealthy owner-drivers when there were only a few thousand cars in the country. A motoring atlas was an aspirational item.

By 1930 the number of privately owned cars had reached the million mark, to double again before the outbreak of war. Petrol maps were seldom given away free in the UK, as they were in the USA, allowing greater competition from mainstream commercial map-makers, automobile associations, tyre manufacturers and other interested parties. Cheap pocket atlases and plans of towns and cities also owe some part of their existence to the needs of the motorist: Philip's ABC Pocket Atlas-Guide series, Bacon's folding maps (ills 72, 73) and the *A1 Atlas* published by Geographia Ltd, a company founded by Hungarian Alexander Gross. Building on this successful format, Gross's daughter Phyllis Pearsall launched the Geographers' A–Z Map Company in 1936, which became one of the most recognisable and mythologised twentieth-century cartographic brands (ill. 74). Such at-

72 (above left)
Bacon's Large Scale Map and Guide to Isle of Wight: For Cyclists and Tourists. Cover of folding map. Published by G. W. Bacon & Co., early 20th century. Private collection

73 (above right)
Bacon's New Half-Inch Maps. Cycling and Motoring: Worcester District. Cover of folding map. Published by G. W. Bacon & Co., early 20th century. The motor car leaves cyclists in its wake. Private collection

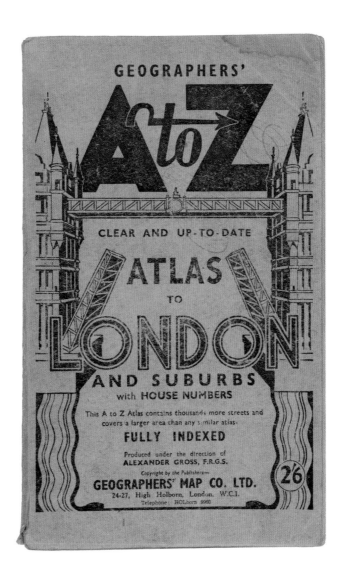

GEOGRAPHERS'

A to Z

CLEAR AND UP-TO-DATE

ATLAS

TO

LONDON

AND SUBURBS

with HOUSE NUMBERS

This A to Z Atlas contains thousands more streets and
covers a larger area than any similar atlas.

FULLY INDEXED

Produced under the direction of
ALEXANDER GROSS, F.R.G.S.

Copyright by the Publishers—
GEOGRAPHERS' MAP CO. LTD.
24-27, High Holborn, London. W.C.I.
Telephone: HOLborn 9960

26

74 (above)
*Geographers' A–Z: Clear and Up-To-
Date Atlas to London and Suburbs.*
Geographers' Map Co. Ltd., c. 1950.
Private collection

lases continued to serve other users too. Author Ron Ramdin emigrated from Trinidad to
England in 1962. On his arrival in London, the Tube map and a pocket street atlas proved
indispensable, and though 'word of mouth was used' the atlas was a 'necessary append-
age' for generations of immigrants, as with any newcomers, guiding them 'in their search
for jobs and housing or shopping and getting around'.[67]

The newest form of transport also contributed to the rediscovery of older forms
of cartography. John Ogilby's road maps inspired a series of decorative maps drawn by
Alfred Taylor for Pratt's (remaining in print when Pratt's was taken over by Esso later in
the decade). Alexander Duckham of Duckhams Oil reprinted the entire Ogilby atlas in
facsimile to mark the company's fortieth anniversary in 1939, though it was not issued
until after the war (ill. 75). In the foreword Duckham mentions that he had hunted for
an original set of the Ogilby maps and 'on this, our 40th birthday, I wish you to accept
as a gift something prized by myself'. Maps were being reprinted for antiquarian interest (such as the facsimile of William Morgan's huge map of Restoration London printed
for the London Topographical Society in 1904[68]) and for their decorative appeal. By the

BRITANNIA

VOL. I

OR an

Illustration

of the Kingdom of

ENGLAND

and dominion of WALES

By a

Geographical or Historical

Description

of the Principal

ROADS.

The Road from
LONDON to BARWICK

IN WESTMOR
LAND
YORK

LIN COLN
SHIRE

mid-twentieth century reprints of antique maps, particularly county maps and world maps by the likes of John Speed and Thomas Moule, had become a popular feature of interior decor.

As Duckham's comment suggests, collecting the originals had also become a popular pastime, spearheaded by a new generation of dealers such as R. V. Tooley and supported by a new generation of carto-bibliographies – a twentieth-century term coined by Sir Herbert George Fordham in 1914[69] – such as Thomas Chubb's *Printed Maps in the Atlases of Great Britain and Ireland 1579–1870* (1927). In time Tooley personally added to the literature of collecting (his *Maps and Map-Makers* was first published in 1949, followed by a number of hand-lists covering specific topics such as the mapping of Africa). Collecting antique maps seems a niche pursuit, a byway of the twentieth-century engagement with maps, but it inspired the confident reappropriation and reinterpretation of the designs as well as the physical maps of the past: Alfred Taylor's petrol maps are the tip of the iceberg.

However, there seems to have been relatively little discussion of the nuanced language of maps in the first part of the twentieth century. A reader consulting the eleventh edition of the *Encyclopaedia Britannica* would have learnt that maps 'are a representation, on a plane and a reduced scale, of part or the whole of the earth's surface'.[70] Over the following pages types of mapping are covered in exhaustive detail, but caution is only advised in terms of potential inaccuracies on the map.

This was strongly reinforced in literature, which gives compelling evidence of what was familiar (or at least natural) for contemporary readers. Where maps appeared in children's fiction, 'X' continued to mark the spot as surely as it did in Stevenson's *Treasure Island* (1883). Fiction for adults was equally literal-minded. Sherlock Holmes sends out to 'Stamford's' (Stanford's) for the latest Ordnance map of that 'portion' of Dartmoor containing Baskerville Hall, as a necessary precursor to leaving for Devon ('my spirit has hovered over it all day. I flatter myself that I could find my way about').[71] Maps adapted from the real world for fictional purposes serve thrillers such as Erskine Childers's *The Riddle of the Sands*[72] and numerous crime novels, such as the map of Galloway created for use with *The Five Red Herrings*, a fiendishly complex plot by Dorothy L. Sayers which relies heavily on train timetables. A map was not required for Sayers's novel *Gaudy Night*, set in an Oxford College: its absence flatters the reader – we are all expected to know the layout of an Oxford College.[73] There is an unwritten rule: the map, where it appears, never lies.[74] This remains true of fantasy maps, such as Tolkien's map of Middle-earth.[75] The reader navigates an imaginary world with a map which, on its own terms, is as reliable as the Ordnance Survey.

The importance of using reliable maps for wayfinding had assumed almost talismanic properties by the early twentieth century. In *The Thirty-Nine Steps*, John Buchan's fugitive hero Richard Hannay, a resourceful amateur hurled into a world of murder and espionage, reaches for an atlas when he needs to find somewhere sparsely populated enough for him to lie low and practise his 'veldcraft'; he memorises a 'map of the neighbourhood' when hiding out in Scotland; and when setting out to round up the nest of German spies, his principal requirements are the loan of a car and a 'good map of the roads'.[76]

This seems to have communicated itself to Alfred Hitchcock, who expanded on cartographic themes and images when he filmed Buchan's novel in 1935. A crumpled map, clutched in the dead hand of a glamorous female agent (Hitchcock's cinematic substitution for 'nervous little' Scudder) is revealed as the environs of the Scottish village of Killin. At a stroke Hitchcock circumvents one of the most unlikely coincidences in Buchan's thriller – how Hannay eventually stumbles upon the foreign spies. The map is quickly overlapped with the image of the dead woman: 'the flashback ... in what might be a shot unique to early sound cinema, turns the cartographic object into a "talking memory-map" in which the voice infuses the areas shown (especially "Killin") with anxiety' (ill. 76).[77] However, the overall purpose is the same as Buchan's original: the reader/viewer is supplied with topographical accuracy.

Hitchcock's creative use of map imagery demonstrates how maps began to appear in other, less tangible forms. Weather maps made their first televised appearance in the earliest days of the BBC, in 1936,[78] and maps have rarely been away from the small screen since, worked into the title sequences of popular shows such as police procedural *Z Cars* and sitcom *Dad's Army*, and frequently accompanying television news reports – becoming increasingly interactive with the presenters in the studio by the end of the century.[79] Digital mapping is also a product of the twentieth century. Created to assist in the production of paper maps, it has become closely associated with virtual technology. Already, by 1975, geographer and map historian J. Brian Harley felt confident in expressing his view that the 'conventional printed map no longer holds a monopoly'.[80] By the 1990s the Ordnance Survey archive was fully digitised. However, the expectation of fast Internet access, and access to GPS mapping through smartphone applications and satnav, are twenty-first-century developments. Only in the last ten to fifteen years have formerly military-grade technologies become an accepted part of everyday life, now so universal that the recent nature of their arrival can easily be forgotten.

The power of maps has perhaps never been in doubt, and the more obvious forms of propaganda fell out of favour early in the century. Metaphorical maps of Europe had been a popular satirical form since the Crimean War in the 1850s; such maps are populated by heroes and grotesques, human and animal, striding and strutting across the Earth's surface or embodied in the map, where they fill the available space, pushing and shoving at the political boundaries that constrain them. They rely on instantly recognisable national stereotypes, often arrayed in the spiked and plumed helmets of the armies as they went to war in August 1914. The maps derive part of their strength from subverting the neutral, scientific, cartographic form so outrageously, and their popularity peaked in the early months of the Great War. Just as abruptly the genre fell out of use: very few were produced after Italy's entry into the war in 1915. As the euphoria that greeted the outbreak of war dissipated, maps like these lost their comic appeal, and the public rapidly came to question and reject the crudest propaganda served up during the first part of the war. Octopuses and spiders slither and crawl over some later cartographic propaganda, but metaphorical maps in general became much darker and, like many travel posters, increasingly influenced by the simple lines of Art Deco and other twentieth-century movements and styles.[81] The slim and sinister Japanese octopus

76
Poster for the 1935 British film,
The 39 Steps, directed by Alfred
Hitchcock, along with a screenshot
from the film capturing a
cartographic flashback

envisaged by Pat Keely on a 1944 propaganda poster intended to boost Dutch morale (ill. 77) is in sharp contrast to the buffoonery of Fred Rose's Boer War-era 'serio-comic map' of Europe, *John Bull and his Friends*, not least in the exchange of a traditional projection for one inspired by air travel.

Persuasive cartography later in the century, even where the maps are simplified, tended to rely upon skilful presentation of features, borders and statistics to make a point; in other words, even where the cartographic imagery is basic it relied on an underlying perception that maps are scientific. Promotional maps appear to offer accuracy wrapped between branded covers but, given the layers of information which of necessity are stripped away, the most innocuous-seeming maps still attempt to lead their audience. Even a humble 1980s *Happy Eater Route Map* attempts to draw the user away from the motorways to any one of its chain of family-friendly and inexpensive roadside eateries on the A-road network.[82] Board games also provide a natural vehicle for persuasive maps of all kinds. Educational and otherwise, cartographic games have a history stretching back at least 300 years and, though by no means exclusively aimed at children, wartime games such as *Black-out*, or the 1950s strategy game *Risk* contributed to the general background chatter of cartographic images which twentieth-century citizens experienced from an early age. In some cases existing maps were repurposed, such as the 1908 UERL system map, which formed the board for *How to Get There* (ill. 78). In its new form the map was deployed to familiarise the Edwardian travelling public with the new deep-level Tube lines, and the 'Underground' brand resulting from a new spirit of cooperation between the UERL and the other independent underground railway companies: on the map, all lines were now granted equal weighting, with the implicit acceptance that passengers changed trains rather than remaining on one line for the duration of the journey.

The proliferation of self-created maps towards the end of the century may have sharpened popular understanding. Where commercial companies would not get involved for monetary, political or other reasons, it became possible for special-interest groups or individuals to print and distribute their own maps. For example, the Man to Man bookshop in Notting Hill published *London Gay to Z* in 1977 (ill. 79), five years before mainstream publisher Bartholomew's licensed the use of its cartography as a base map for a Spartacus guide to gay London.[83] By the 1980s the Campaign for Nuclear Disarmament had produced its own detailed operations maps, for the purpose of direct action against convoys carrying nuclear weapons.[84] Maps have frequently been employed as souvenirs – as fitting reminders of great state occasions, of coronations or victory parades, but as the century progressed these were also commissioned by private individuals or minority groups. During the miners' strike of 1984–5 regional branches of the National Union of Mineworkers commissioned ceramic plates which were sold in aid of miners' welfare, some of which feature maps of mining districts, with the local NUM headquarters marked in red (ill. 80). The miners' strike has been absorbed into the history of radical protest, which in modern times is rarely associated with souvenir china, but this overlooks the fact that the miners were working people first and protestors second; they wished to maintain their jobs and way of life. The plates are propaganda of a sort,

77

Pat Keely, *Indie Moet Vrij! Werkt en Vecht Ervoor!* James Haworth & Brothers Ltd., 1944. Maps CC.6.a.77

INDIE MOET VRIJ!

WERKT EN VECHT ERVOOR!

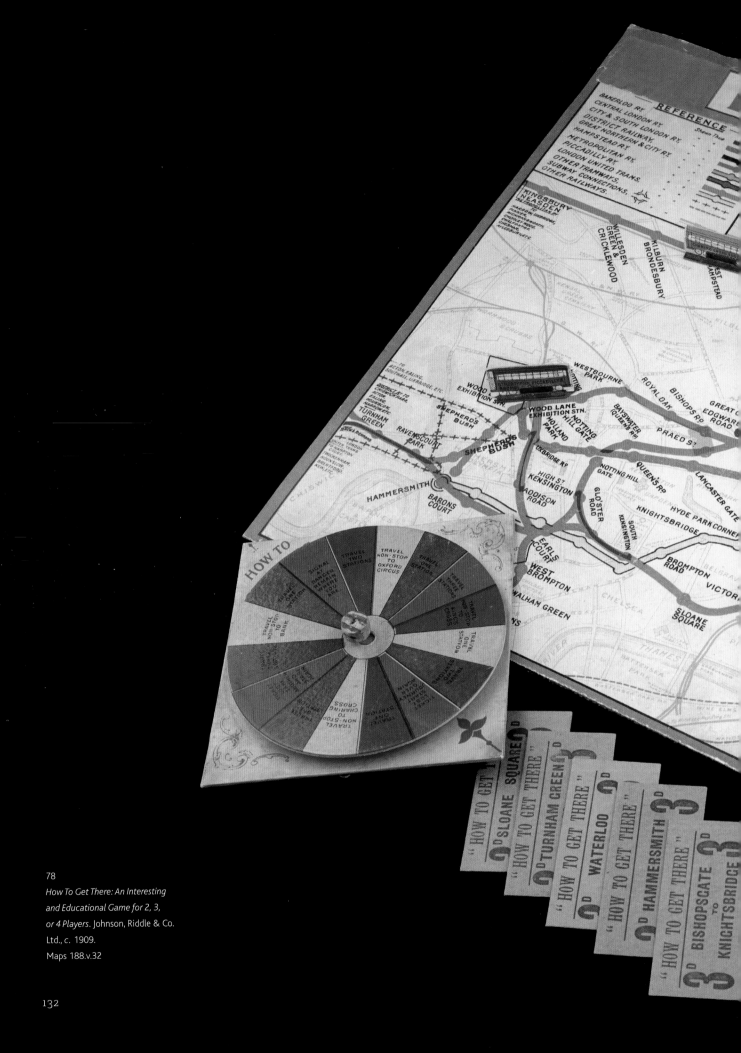

*How To Get There: An Interesting
and Educational Game for 2, 3,
or 4 Players*. Johnson, Riddle & Co.
Ltd., c. 1909.
Maps 188.v.32

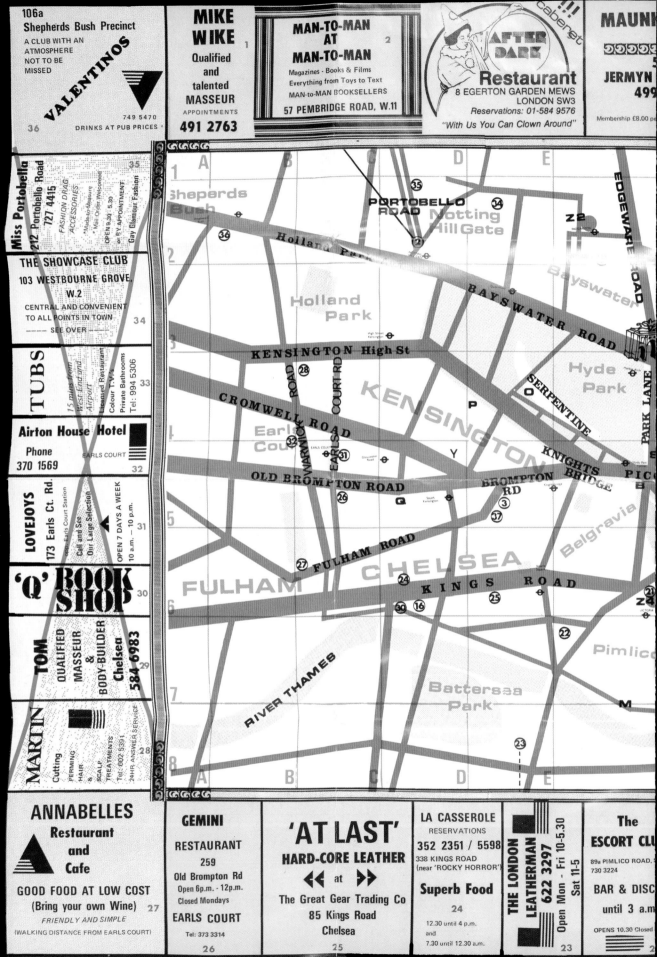

PLACES OF INTEREST

A PICCADILLY CIRCUS
B HYDE PARK CORNER
C BUCKINGHAM PALACE
D ST. JAMES PALACE
E ROYAL ACADEMY
F SHAFTESBURY AVE
G CHARING CROSS RD
H NATIONAL GALLERY
I COVENT GARDEN
J OPERA HOUSE
K TRAFALGAR SQUARE
L PARLIAMENT + ABBEY
M TATE GALLERY
N CARNABY STREET
O LIDO (Swimming)
P ALBERT HALL
Q HARRODS
R AMERICAN EMBASSY
S HILTON HOTEL
T TOWER OF LONDON
U TOWER BRIDGE
V FESTIVAL HALL
W NATIONAL THEATRE
X St PAULS. CATHEDRAL
Y MUSEUMS (S. Ken)
Z See "STATION's KEY"
BM BRITISH MUSEUM
PO POST OFFICE TOWER

KEY TO STATIONS ●
Z1 WATERLOO (For South-West)
Z2 PADDINGTON (For West)
Z3 MARYLEBONE (Local)
Z4 VICTORIA (For S.E. & Abroad)
Z5 KINGS CROSS (For North)
Z6 EUSTON (For N.W.)
Z7 St. PANCRAS (Midlands & Local)
Z8 CHARING CROSS (For S.E. & Abroad)
Z9 LIVERPOOL STREET (For East & Abroad)

79 (previous page)
London Gay to Z: Classified
Directory with Maps. Published by
the Man to Man Bookshop,
Notting Hill, London, 1977.
Private collection

80 (above)
Commemorative ceramic plate
by the National Coal Board
showing the Lancashire Coalfield,
1980s. Despite their political
differences, plates designed for
the National Coal Board and the
National Union of Mineworkers
were stylistically similar.
Maps 188.v.38

81 (above)
Chris M. Dickson, ASCII map of
Great Britain, 1998

certainly, but underpinned by a more subtle desire to record and commemorate one's own past. The proud history of the local pits is encapsulated in cartography and finds permanence in ceramic work.

At the other extreme, ASCII (American Standard Code for Information Interchange) maps enabled anyone with a computer to create their own maps using repeated punctuation marks and dingbats, with 'variations in greytone density achieved by overprinting characters on a black and white printer' (ill. 81).[85] Photocopiers had become widely available by the 1970s, with the first electrostatic colour copier released by Xerox in 1973. This gave complete flexibility over print runs and content (though not necessarily without consequences in terms of copyright infringement), and photocopied maps could be made and used by groups as diverse as birdwatchers[86] and political protestors. As it became clear that maps could as easily be created from below as from the top down, a body of literature on the potential of maps to deceive began to accrue. Whether or not that has affected the majority view of maps remains to be seen. The desire to be on the map, or even at the centre of it, is of long standing and has never been so readily achievable. We are now always at the centre of our own digital maps: a mini-map is trained on anyone with a smartphone.

A broad degree of map-mindedness was established before 1900, but over the course of the twentieth century it mushroomed until it became all-encompassing. Much of the available written evidence concerns the intentions of educators, and the pastimes of the tolerably well-off. It is notoriously difficult to explore the day-to-day lives of the majority of the population; it is seldom recorded or considered worthy of record, hence our reliance on popular literature. But the best arbiter is perhaps the twentieth-century map archive itself, where we began. Millions of maps were considered worth manufacturing, to be sold cheaply or given away in vast numbers, to inform, entertain or persuade, leaving a rich seam for analysis – much of it surviving quite by chance. Map imagery had become instantly recognisable and, from Edwardian postcards to the Live Aid logo, maps reached into every part of daily life.

They shall beat their swords into plowshares, and their spears into pruninghooks: nation shall not lift up sword against nation, neither shall they learn war any more.

ISAIAH

B.C. 760-698

The only excuse for war is that we may live in peace unharmed.

CICERO

B.C. 106-43

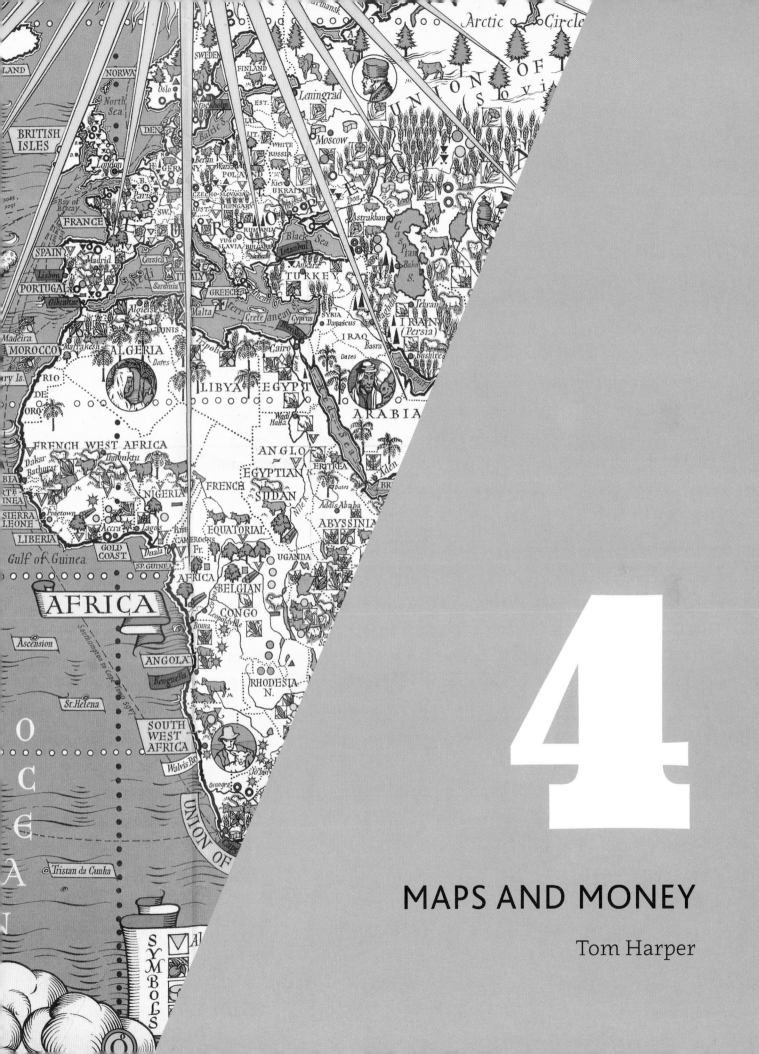

4

MAPS AND MONEY

Tom Harper

28

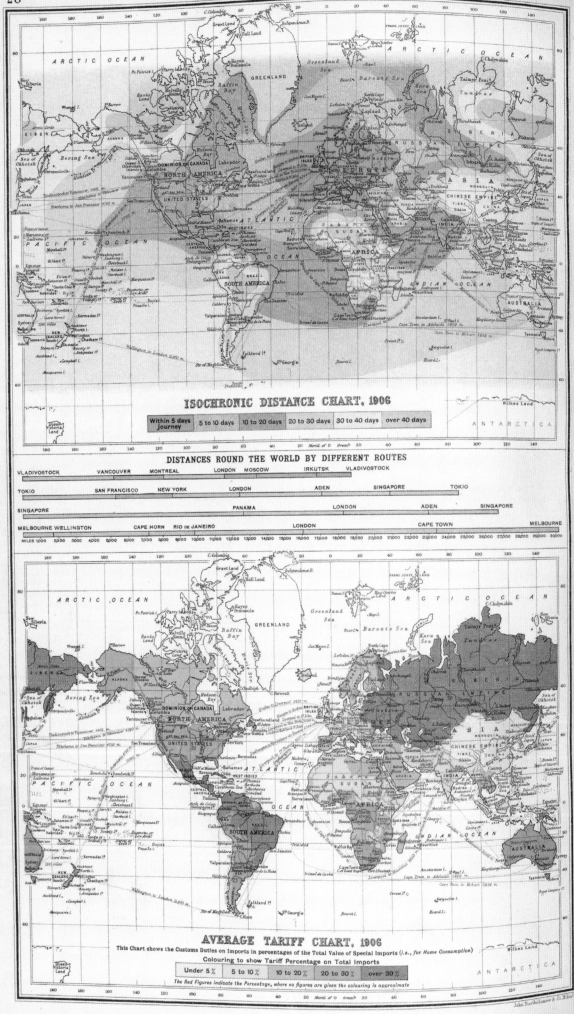

ISOCHRONIC DISTANCE CHART, 1906

Within 5 days Journey	5 to 10 days	10 to 20 days	20 to 30 days	30 to 40 days	over 40 days

DISTANCES ROUND THE WORLD BY DIFFERENT ROUTES

| VLADIVOSTOCK | VANCOUVER | MONTREAL | LONDON MOSCOW | IRKUTSK | VLADIVOSTOCK |

| TOKIO | SAN FRANCISCO | NEW YORK | LONDON | ADEN | SINGAPORE | TOKIO |

| SINGAPORE | PANAMA | LONDON | ADEN | SINGAPORE |

| MELBOURNE WELLINGTON | CAPE HORN | RIO DE JANEIRO | LONDON | CAPE TOWN | MELBOURNE |

MILES 1,000 2,000 3,000 4,000 5,000 6,000 7,000 8,000 9,000 10,000 11,000 12,000 13,000 14,000 15,000 16,000 17,000 18,000 19,000 20,000 21,000 22,000 23,000 24,000 25,000 26,000 27,000 28,000 29,000 30,000

AVERAGE TARIFF CHART, 1906

This Chart shows the Customs Duties on Imports in percentages of the Total Value of Special Imports (i.e., for Home Consumption)

Colouring to show Tariff Percentage on Total Imports

Under 5 %	5 to 10 %	10 to 20 %	20 to 30 %	over 30 %

The Red Figures indicate the Percentage, where no figures are given the colouring is approximate

John Bartholomew & Co. Edin

ollowing an interpretation of Marxist theory, the key narratives of the twentieth century were economic in origin.[1] The financial, industrial and commercial factors that had already accounted for the social and political patterns apparent in 1900 went on to drive the changes of the ensuing century through war, revolution and social transformation. The centrality of the economic dimension in the twentieth century can be gauged in world maps produced throughout the period. In John Bartholomew's *Atlas of the World's Commerce* of 1907 (ill. 82), as in the *Oxford Economic Atlas of the World* of 1972, economic data held centre stage, with the locations of commodities and raw materials and the sites of industry shown linked to the global economy by means of railways and shipping routes. The patterns and concentrations of world trade were embedded in the physical and political geography which made up the world's cartographic reality.

Maps and graphs are commonly used in literature on the economic history of the twentieth century to support discussion on the locations of resources and the flow of trade. But less has been made of the more active role that maps played in shaping the economic history of the twentieth century. Maps were not only commodities in themselves but supported capitalism both practically and ideologically. Redressing that balance of understanding is the aim of what follows.

First I will explain how maps were produced as commodities to an even greater extent in the twentieth century than they had been in previous periods. Map production was a profit-making exercise, and as the margin of profit increased, the cost of actually producing the maps decreased. This was due to an increasing consumer base on the one hand, and the application of more efficient data gathering, production and marketing processes on the other. Next I will demonstrate how, having been sold to the public and businesses, maps were then used in a capitalist context to create more profit. This profit was made through the practical capabilities of maps, for example mining maps, or business directory maps, which actively supported commercial activities. Profit was also generated through the ideological use of maps. Here, the message of economic vitality that was contained in cartographic images of the economic geography of the

82
J. Bartholomew (ed.), 'Isochronic Distance Chart, 1906/Average Tariff Chart, 1906.' *Atlas of the World's Commerce.* G. Newnes, 1907.
Maps 48.e.9

world reinforced and perpetuated the reality of that message in the mind of the viewer.

Maps underpinned the ideological basis for the capitalist world, an ideology which proved resilient throughout the economic uncertainties of the twentieth century. However, at the end of the 1960s maps began to appear that had not been produced for profit, and which did not set out to make a profit. These maps were produced to criticise the capitalist world, which by the 1970s was seen by a portion of society as having been the cause of inequality and environmental negligence.

In describing these three aspects of the twentieth-century story of maps and economics I shall refer to four main periods of the century. First, the time up to 1918 witnessed the high point of the imperial period, a climax of the European-led industrial capitalist system of the previous century in which much of the landscape of the twentieth century, including the emergence of maps as capitalist tools, was formed. Second, following the cataclysm of the First World War, the interwar period of 1919 to 1939 saw periods of economic boom and the world's first global financial crash and economic slump, with recovery largely a consequence of rearmament. Third, following six years of war, the period from 1950 to 1972 was a 'golden age' of prosperity, increased consumerism and industrial development, alongside decolonisation of the developing world. Finally, the disorienting period of 1973 to 2000 contained the apparent triumph of capitalism, the fall of communism and growing economic uncertainty.

I use the term 'economic' here in reference to everything that has to do with the production, distribution, and sales and consumption of goods and services. By 'capital' and 'capitalist' I mean the economic system which supplanted feudalism in much of Western Europe by the early sixteenth century, and by which a measure of profit is extracted by a private business owner from the working class.

1. Map production in a capitalist context

Maps had been produced for profit for many centuries prior to 1900. The introduction of printing to Europe during the fifteenth century, the expansion in travel and mercantile trade, and the value of land all contributed to the developing trade in geographical products such as maps and atlases. They were made and sold by specialist networks of mapmakers and map publishers in parts of Europe, Asia and eventually North America.[2] During the second half of the nineteenth century maps became commonplace objects in Western society, and at the peak of European imperial power in 1900 maps were an economic commodity and mapmaking a profitable business. I look in turn at the data compilation, production, sale and marketing of maps during the twentieth century, as a preface to a discussion of their exploitation and use for capitalist ends.

Data gathering
Data gathering as the traditional way of compiling geographical information was gradually supplanted by more expedient methods during the twentieth century. Land surveying, the process by which a surveyor or team of surveyors mapped land through

83
R.A.F. aerial photo-mosaic of the area around Prestwick Airfield, near Glasgow. Ordnance Survey, 1946. Maps OSM 26/32 NE

MONKTON

ADAMTON

SPRINGBANK

PRESTWICK

Approximate Scale 1:10,560 or about 6 inches to 1 mile

Metres
1,000

0

1,000

2,000

3,000

Metres
4,000

Feet
1,000   0   1,000   2,000   3,000   4,000   5,000   6,000   7,000   8,000   9,000   10,000   11,000   12,000   13,000   14,000   Feet

Published by the Director General, Ordnance Survey Office, Chessington, Surrey.

Prepared from Air Photographs taken by R.A.F. May, 1946.

direct contact with it, capturing data through measuring devices such as a chain and theodolite, had originated in the sixteenth century and continued as a local practice even after 1900. During the First World War photographs began to be taken of the ground below from aeroplanes, and the information was used to compile and update maps.[3] Aerial photographic survey became more streamlined as equipment and techniques improved, largely superseding the laborious and expensive method of surveying large areas by ground survey (ill. 83).

Aerial survey was joined after 1945 by surveying through remotely sensed map imagery. This method of data gathering used the information in photographs taken by satellites orbiting the Earth. Government-owned satellites such as Landsat (1972–) and commercially owned ones such as SPOT (1986–) were in orbit around the Earth, and so capable of repeated data capture.[4] The more comprehensive vision offered by the photographs was enhanced by an improvement in their quality from the 1960s, while developments in computer science increased the ways in which their data could be interpreted and used. All surveying techniques described above were used in the twentieth century, some being more appropriate than others for particular areas. For example, small areas continued to be served by traditional methods.

Production

Reducing production costs made maps more affordable for greater numbers of people during the first decade of the twentieth century. Maps became cheaper to produce, and this was reflected to some degree in their cost to the purchaser. One contributing factor was the introduction of more expedient photo-mechanical production and printing processes into the commercial mapmaking sphere. Photolithography and offset lithography had been developed by the military (thanks to huge investment) in the previous century. Expensive printing materials such as copper, in use since the fifteenth century, had been replaced by stone during the nineteenth century, while the photo-mechanical image transfer process replaced the hand of the artist-engraver.[5]

From the 1960s digital production processes also reduced production costs. The first maps drawn from computer software, however, were not of a high visual quality, and it was only when the pen plotter was introduced in the 1970s that the quality of the map image improved – though ironically, it mimicked the traditional hand-drawn style.[6] In other ways the computer revolution of the 1960s affected the map industry more profoundly than the transition to lithography circa 1870. It meant that maps could be stored as digital files and only printed when necessary, and mapping companies were no longer required to update and republish new maps whenever their old ones were superseded. Individual sheets could be printed on demand, and output tailored to accommodate the needs of specific users, such as the tourist market or the mining industry. By the twenty-first century, with the emergence of computerised hand-held devices, it had dawned on map producers that maps did not necessarily need to be taken off the screen at all.

84
Ordnance Survey Road Map of Great Britain. Cover proof. Ordnance Survey, 1936. Maps CC.5.b.31

85 (left)
Ordnance Survey Tourist Map of the Lake District. Cover proof. Illustration: Ellis Martin, 'Derwentwater from Skiddaw', c. 1920.
Maps CC.5.b.31

86 (right)
Ordnance Survey Town Map:
Leicester. Cover proof.
Ordnance Survey, 1932.
Maps CC.5.b.31

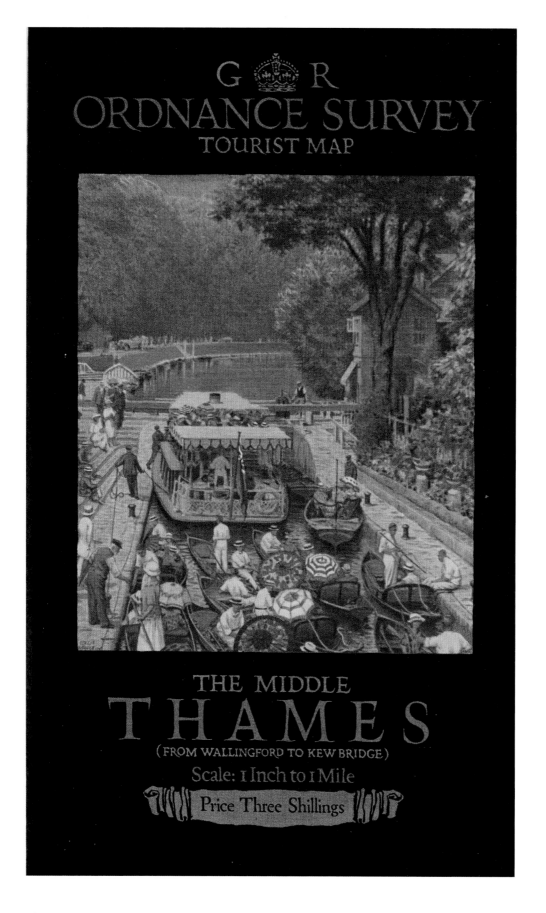

Marketing

Successful twentieth-century marketing techniques contributed to greater sales of maps. Adverts announcing the publication of new maps had occurred in news-sheets in Britain from the sixteenth century.[7] By the twentieth century this form of publicity had increased still further, as map producers were able to market their map products for specific audiences. As a result of the compulsory teaching of geography in schools, which necessitated up-to-date cartographic products, marketing to schools became particularly focused in Britain after 1900, and the figures provide evidence for this. Compared with the period from 1881 to 1900, where 147 atlases and maps with 'school' in the title were published in the UK, 222 were published between 1901 and 1920. The number produced between 1921 and 1940 was a third higher again.[8] Packaged 'education courses' consisting of atlases, textbooks, globes and wall maps were expertly marketed by the likes of publishers George Philip and Edward Stanford to school education boards.

Maps were more strongly marketed for leisure purposes during the interwar period. In the 1920s the Ordnance Survey of Great Britain began to produce tourist maps for the leisure market, with evocative illustrated covers and in a convenient folded format (ill. 85). These maps were marketed to walkers and to the burgeoning numbers of motor car owners. Other maps, such as the town plans of the Geographers' A–Z Map Company (from 1911), employed a convenient and modern format with elegant design in order to appeal to more users, particularly after 1945. During this period, maps and travel guides were marketed as essential accompaniments to holidays.

We can see how changes in the production processes of maps made it more profitable to sell them in the twentieth century. They were also cheaper to produce, and people were more able to afford them than ever before: on average, wages rose in Britain during the twentieth century. John Bartholomew's half-inch folded map of Great Britain was priced at the equivalent of £2.85 in today's money at the beginning of the century. In 1920 it cost £1.20, under half as much. In 1970 it was almost a third less in relative terms.[9] From the 1970s, however, the price rose again in line with a more general increase in the price of consumer goods due to inflation.

2. Map use in a capitalist context

The capitalist functions of maps did not cease once they had been sold. Maps were sold to a growing business as well as a public audience during the twentieth century. For example, the twentieth-century records of the map publisher George Philip show a wide range of business customers for map services, including Cambridge University Press, British Rail, Cunard, ESSO and Marconi.[10] Maps were exploited by commercial companies and other businesses for various profit-making activities, thanks to the subtle ways in which they were used in public and social as well as business contexts.

87
Ordnance Survey Middle Thames Tourist Map. Cover proof. Ordnance Survey, 1923.
Maps CC.5.b.31

Geographical Section. General Staff. Nº 3925.ᴮ
Published at the War Office. 1932.

TRADE ROUTE CHART.

The object of this Chart is to show the annual value of the British trade on the various trade routes and sections thereof, or passing important points.

All figures represent £ million and are based on an average of the years 1927-1929.

The trade which has been identified for the purpose of this chart amounts to £3252·9 million, or 98·6 % of the "Total Trade of the British Empire" as shown on p. 88 of the Statistical Abstract of the British Empire. Of the identified trade £129·4 million is known or assumed to be carried on across land frontiers (including the Great Lakes of North America); the remainder, £3123·5 million, is known or assumed to be by sea.

The red tracks represent, roughly, the figures followed by identified trade and the figures against them show the value of trade passing annually along each section. The blue lines represent steamer tracks for which figures cannot be shown in detail. In general it has not been possible to show separately the value of the trade on tracks to individual ports in each country.

The figures under the name of each foreign country represent the total identified trade of the British Empire with that country. Figures under the names of British Empire countries show the total identified foreign trade of that country, as shown on the chart.

In some instances it has not been possible to ascertain definitely the division of trade between the several routes serving a particular country. In such instances the trade has been divided on an estimate based on the data available. Although exact arithmetical accuracy cannot be guaranteed, it is believed that the general picture given of the trade in each area is substantially correct.

For purposes of detailed study, the use of the Chart should be supplemented by reference to the tables from which it has been compiled.

A.—*Trade of the British Empire by Countries*: Showing the total individual trade of each Empire country with each Empire and foreign country.

B.—*Note on division of trade by Routes and Seaboards*: Showing the method by which trade has been allocated in the case of countries having more than one seaboard or served by more than one route.

C.—*Key Figures and Area Summaries*: Showing the total trade in important areas or passing important points, with its general composition.

D.—*Trade by Routes and Areas*: Showing in detail the composition of the trade on each Route and each section of each Route.

E.—*Imports and Exports of the U.K., British Dominions and Colonies*: Showing the trade of each Empire country with each Empire and foreign country, distinguishing Imports and Exports, and showing totals (a) for trade with U.K.; (b) for trade with rest of Empire; (c) for trade with foreign countries.

N.B.—For explanation of small discrepancies between totals in this table and those in table "A", *see* note to table "A".

The scale of the Chart rendered it impossible to show in detail the trade of tracks in the following areas :—

Waters surrounding the British Isles: For details, *see* table "C", p. 3.
Mediterranean Sea: For details, *see* table "C", p. 7, and table "D", pp. 10-12.

N.B.—The figures on the chart show the total traffic in each area of the Mediterranean.

India and Ceylon Area: For details, *see* table "C", p. 8.

THE BRITISH EMPIRE

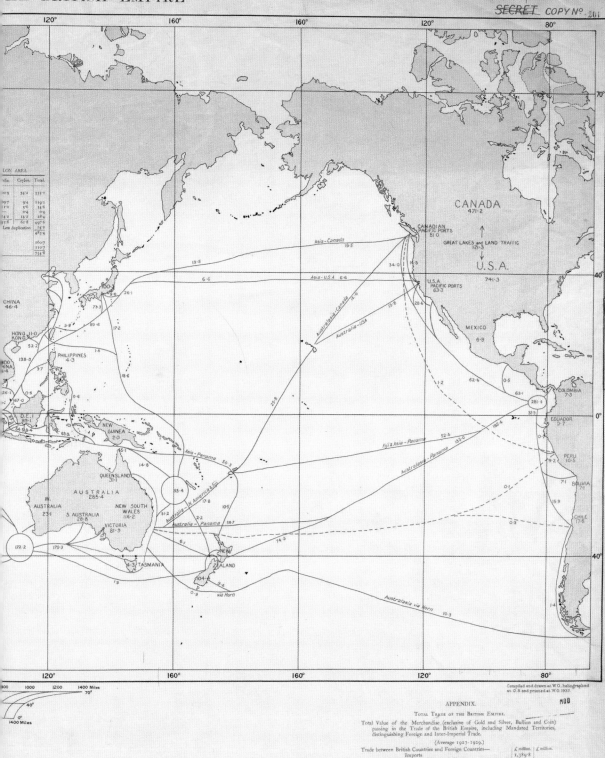

Compiled and drawn at W.O. heliographed
at O.S. and printed at W.O. 1932.

M08

APPENDIX.

TOTAL TRADE OF THE BRITISH EMPIRE.

Total Value of the Merchandise (exclusive of Gold and Silver, Bullion and Coin) passing in the Trade of the British Empire, including Mandated Territories, distinguishing Foreign and Inter-Imperial Trade.

(Average 1927-1929.)

	£ million.	£ million.
Trade between British Countries and Foreign Countries—		
Imports	1,389·8	
Exports of Domestic Produce	920·0	
Re-Exports	133·7	
Total Trade with Foreign Countries		2,443·5
Trade between the United Kingdom and other British Countries—		
Imports (a)	372·6	
Exports of Domestic Produce	326·3	
Re-Exports	22·6	
Trade between British Countries other than the United Kingdom (b)	135·8	
Total Inter-Imperial Trade		857·4
Grand total		3,300·9
Trade with Foreign Countries		74·0 %
Inter-Imperial Trade		26·0 %

(a) Includes value of diamonds imported from British S. Africa and from the Territory of S.W. Africa, not included in United Kingdom Returns.

(b) "For the purpose of ascertaining the 'Total Inter-Imperial Trade' of the British Empire, note has been taken of Imports only, as the goods imported into one country appear in the return of goods exported from another country."

N.B.—In the trade returns of each Empire Country individually, the Inter-Imperial trade appears as a much larger proportion of the total trade, Imports and Exports being both taken into account.

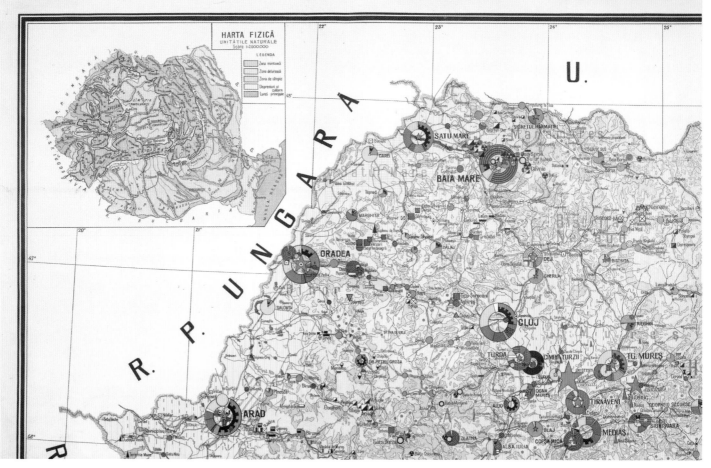

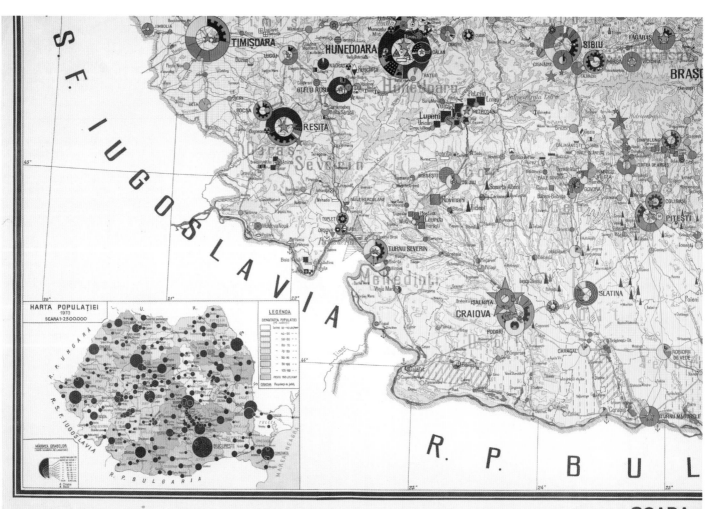

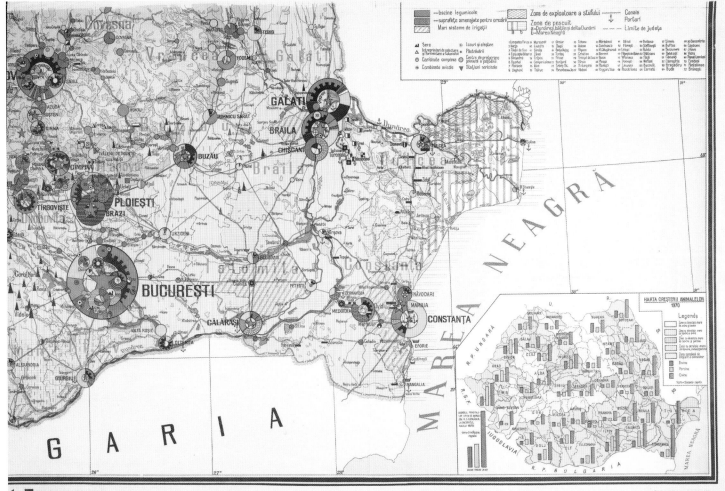

THE PRACTICAL USE OF MAPS FOR PROFIT

During the twentieth century, as in previous and later periods, maps were sold to specific business users who incorporated them into their commercial, profit-making activities. These maps were often highly specific and unembellished, and contained complex and sensitive information for use by a select audience.

Military and naval maps

These made possible and protected the nineteenth-century imperial world. Continuing in the twentieth century, Britain's empire links were maintained by naval control of the seas, which was in turn supported by hydrographical knowledge contained in charts. Military topographical maps were deployed in the acquisition of trading posts and their defence, and in protecting the flow of trade back to the metropolitan heart of empire by sea or rail (ill. 88). Maps were used to plan and realise the construction of vast railways in Siberia, Canada and Africa, and underwater telegraph cables along which fresh forms of finance were communicated. Restricted military and naval mapping, including satellite imagery, protected international trade and preserved the resultant geographical economic reality.

Academic and state economic maps

Produced throughout the twentieth century by geographical and scientific societies, by higher education establishments and by government departments, their purpose was to present economic geography with a high level of comprehensiveness and accuracy so that business decisions could be made based on them. For example, the *National Atlas of Sverige* of 1957 presented the economic geography of Sweden in such comprehensive terms that business decisions on timber and fishing quotas could be made from it. Elsewhere, the *Harta Geografico-Economica* by the Romanian Institute of Geography of 1970 gave the location and concentration of resource and industry in such a way that it was possible to predict further business outcomes based on the picture (ill. 89). The *National Atlas of the Democratic Republic of Afghanistan* was produced during the Soviet occupation of the country in the early 1980s by the Polish map company Geokart. Its purpose was to initiate the practical exploitation of the occupied and backward yet resource-rich territory.

The European Economic Community (1957, later the European Union) produced maps for practical use by its members. Its *Regional Development Atlas*, produced from 1978 by a team of economic advisors, analysts and geographers, aimed to analyse and identify the potential for economic growth and the diversification of trades. It focused particularly on post-industrial areas of Europe such as Wallonia and north-east England, which had been devastated by war and international competition. Individual states also produced maps to assist economic development. For example, the UK government Wales Office's *Less Favoured Areas* map of 1984 identified deprived areas of the country suitable for economic regeneration projects (ill. 91). Meanwhile, the *Development Projects, Western Australia* map of 1987 showcased a profitable region ripe for further industrial exploitation (ill. 90).

88 (pages 150–151)
Trade Route Chart of the British Empire. War Office, General Staff, Geographical Section, 1932. Maps MOD GSGS 3925a

89 (previous page)
Harta Geografico-Economică. Institul de Geografie al Academiei R.S. România, 1972. Maps 43860.(21)

90 (right)
Development Projects, Western Australia. Department of Land Administration, 1987. Maps X.424

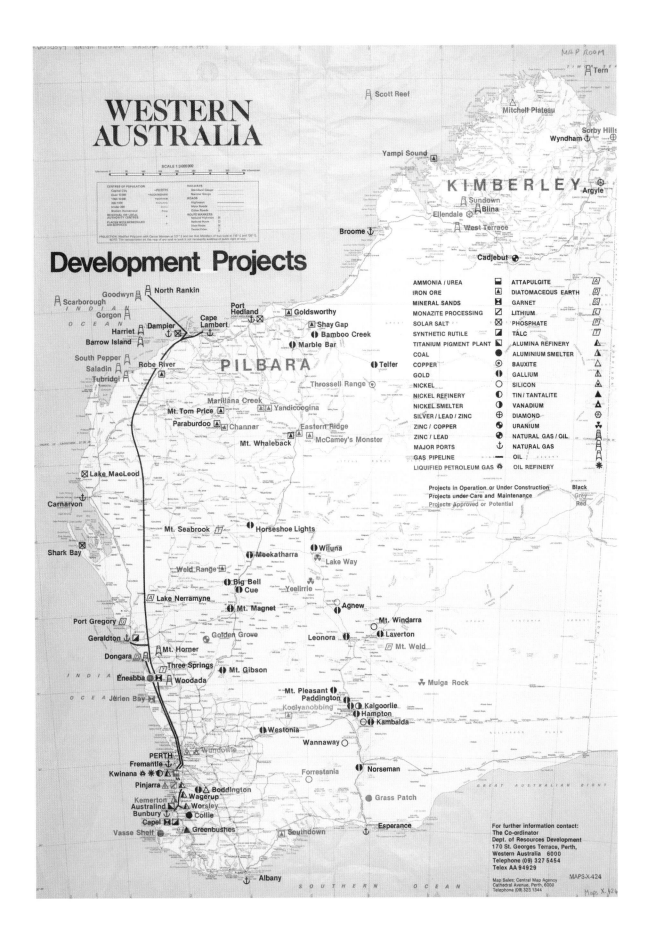

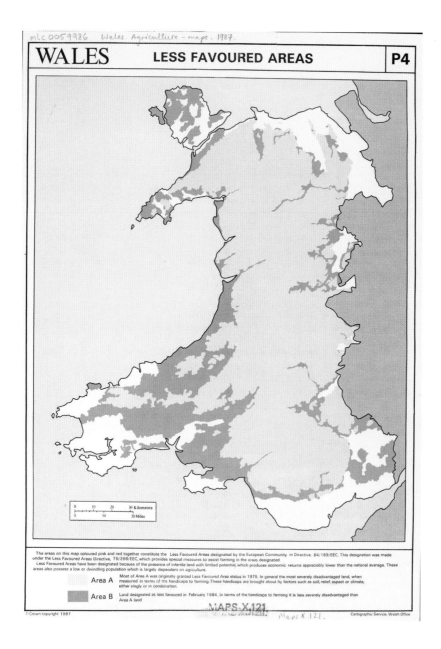

mlc 0059986 Wales. Agriculture - maps. 1987.

WALES   LESS FAVOURED AREAS   P4

The areas on this map coloured pink and red together constitute the Less Favoured Areas designated by the European Community in Directive, 84/169/EEC. This designation was made under the Less Favoured Areas Directive, 75/268/EEC, which provides special measures to assist farming in the areas designated.
Less Favoured Areas have been designated because of the presence of infertile land with limited potential, which produces economic returns appreciably lower than the national average. These areas also possess a low or dwindling population which is largely dependent on agriculture.

Area A — Most of Area A was originally granted Less Favoured Area status in 1975. In general the most severely disadvantaged land, when measured in terms of the handicaps to farming. These handicaps are brought about by factors such as soil, relief, aspect or climate, either singly or in combination.

Area B — Land designated as less favoured in February 1984. In terms of the handicaps to farming it is less severely disadvantaged than Area A land

© Crown copyright 1987   MAPS X.121.   Maps X 121.   Cartographic Service, Welsh Office

91 (left)
Wales: Less Favoured Areas.
Cartographic Services, Welsh
Office, 1984.
Maps X.117

Geological mineral maps

Along with surveying, these were already in their maturity by 1900, with the location
and extraction of raw materials and fossil fuels having long before provided the platform
for the Industrial Revolution in Europe and North America. State-financed geological mapping continued to support processes for the consumption or export of state
resources during the twentieth century. Exploration and prospecting were managed by
state departments such as the Canadian Department of Mines, which licensed concession areas to state-run or private companies with the help of maps. The British War
Office was engaged in mapping the mineral resources of the East Africa Protectorate
in 1904 (ill. 92), which informed lucrative licence agreements. Continental geological mapping and maps showing the locations of mineral resources were produced

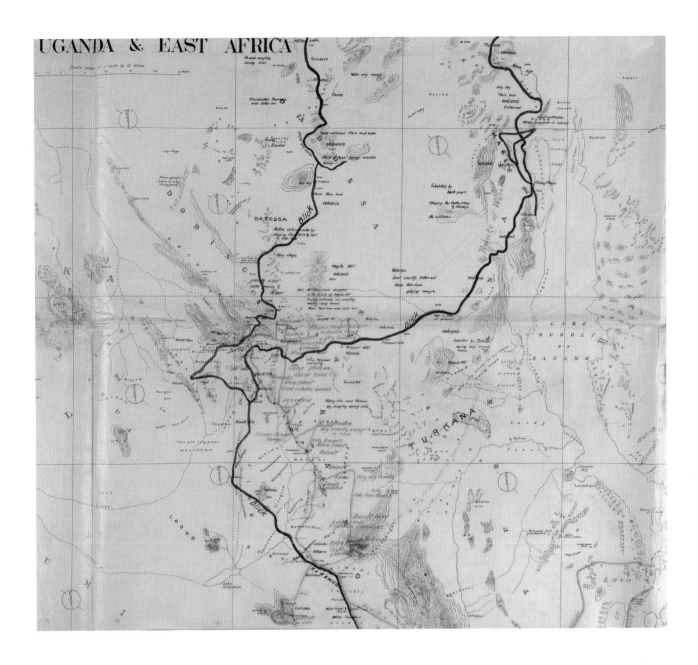

92 (above)
Uganda & East Africa (detail).
British War Office manuscript
map of the protectorate,
showing minerals and resources.
East Africa Syndicate Ltd., 1904.
WOMAT AFR BEA/102

by international organisations such as the United Nations Economic Development
Committee after 1945. Their aim was to provide economic support to newly
independent states.

At a smaller scale, maps assisted the exploitation of individual mines. For example,
highly detailed scientific maps of metal mining areas in Victoria, Australia in 1909
illustrated the limits of mine shafts extending vertically and horizontally through beds,
under townships and against abutting concession areas (ill. 93). Similar maps of the
Michigan copper mines had a practical, planning and administrative function for an
industry which fuelled the rapidly growing North American car industry. Prospecting in
the latter area intensified during the two world wars to fuel the war machine.

Oil and gas concession maps

Used in the management of licence areas for oil-rich parts of the world, these superimposed a grid onto a map of a particular area, which acted as a guide for selection and administration. Further maps were produced following the twentieth-century discoveries of gas and oil in the Middle East (1908) and South America (1914). Those of the Middle East reflected in their number the increasing post-1945 Western political and commercial interest in the region. Some of the most distinctive maps were produced by Texas-based oil-trade publication *The Oil Forum*. Its 'Map of Near and Middle East Oil' of 1960 showed a complex scenario of the large zones of overlapping interest for industry members (ill. 94).

The sizes of the oil and gas concession areas of the Middle East were vast compared to the tiny cells promoted for exploration by the Dutch, Norwegian and UK governments for North Sea oil and gas, which had been discovered in the 1950s. The smaller the area, the greater the level of government control over the licensing areas and revenue. Northern European governments became desperate to find an alternative to Middle East oil in the wake of the Organisation of Petroleum Exporting Countries (OPEC) embargoes of 1973 and later. This led to greater interest in the area, and difficult prospecting and drilling operations in the North Atlantic Ocean and North Sea. Increased demand for natural resources drove greater investment in prospecting and research, in which maps played various practical roles. For example, after oil was first extracted from within the Arctic Circle in 1969 more advanced surveying techniques were used to map the Arctic seabed geology and to advance territorial claims over it.

Land use mapping

This emerged on a national scale during the twentieth century because of a desire for a more uniform state administration and control of national land assets. Land use mapping involved categorising individual parcels of land according to how it was used – from grazing and the types of crops grown, to urban development and types of industry. Given the scale and expense of such an operation in vast countries such as Canada, state support was essential. Yet even in smaller places such as the UK, such projects could not be purely commercially led. There the Ordnance Survey provided a large number of sheets after 1911 for the purpose of land valuation and tax purposes, while the Land Utilisation Survey of Great Britain of the early 1930s reflected a popular desire for systematic planning and control of resources. Information on the yield of land was critical during the Second World War because of the threat of military blockade from Germany and food shortages (ill. 95).

After 1945 land use surveys, like geological surveys, were supplied by international organisations to motivate and assist poor and largely agrarian economies. They made use of aerial photography and satellite imagery. Remotely sensed images from the US Landsat programme were formatted and stitched to provide a seductively complete economic (and environmental) picture of large and inhospitable areas.[11] Meanwhile land use changed profoundly during the twentieth century. The economic global trend of the twentieth century saw humanity change from a predominantly rural and agricultural

93

'Sketch map of Alluvial and Deep Leads Systems Bendigo-Huntly, Campaspe, Malmsbury, Kyneton, Trentham, &c.' (detail), from Stanley Hunter, *Memoirs of the Geological Survey of Victoria*. J. Kemp, 1909. C.S.G.356/4

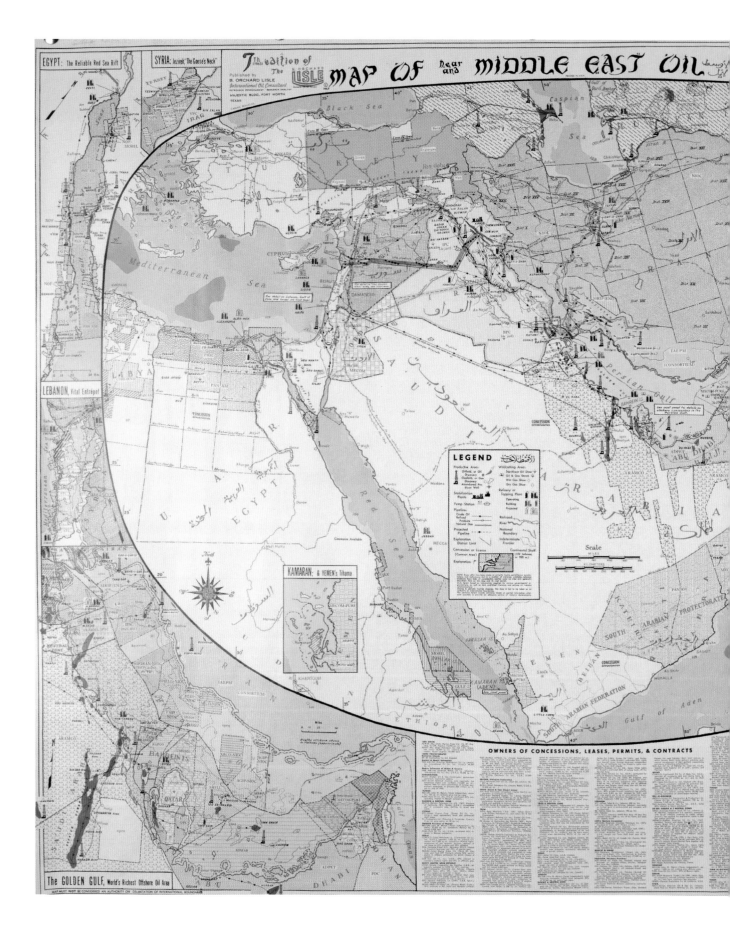

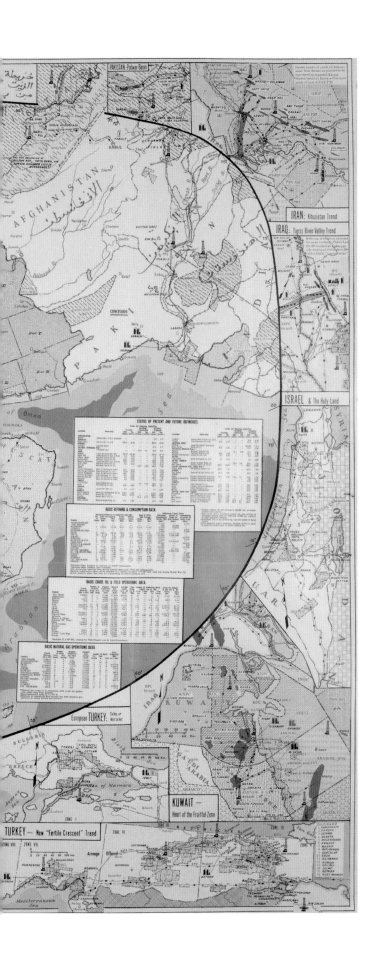

94
'Map of Near & Middle East Oil',
The Oil Forum, 6th ed. B. Orchard
Lisle, 1960.
Maps 46825.(3)

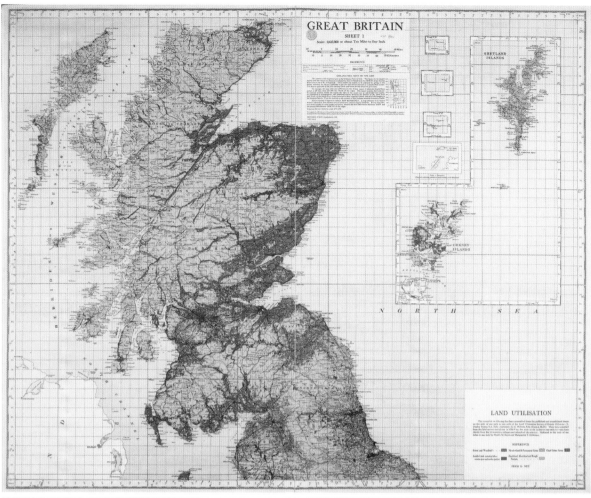

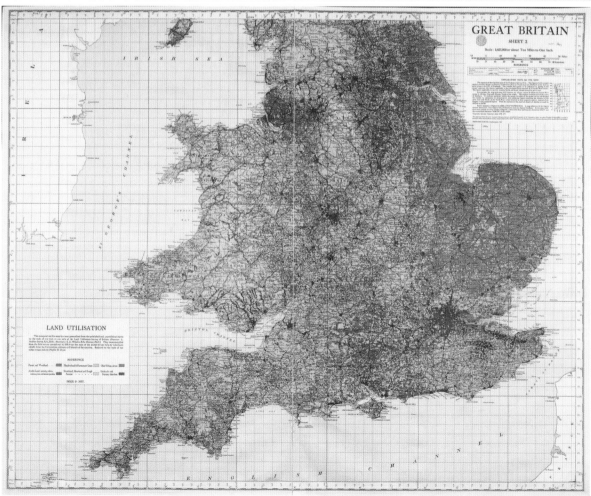

race to an urban, industrial one.[12] Forced farm collectivisation initiated a flight from the land in the Soviet Union from the 1920s, while mechanisation, imports and 'conclusive industrialisation' emptied the farmland of America.[13] Such patterns of economic change were monitored by official maps.

Property maps

These continued to be used as part of the sale and purchase of land. For example, maps were drawn and attached to sales particulars for country estates and parcels of land. In Britain much land was put up for sale after the First World War as a consequence of the decline of the aristocracy, and to make way for urban development. Maps were heavily used as part of the consolidation of new state territory, such as Canada's during the early twentieth century, with maps of boxed and gridded lots drawn up and presented to migrant landowners as part of their land deeds. Such maps revealed the dimensions of the land but, often disappointingly for its new farmer owners, not its quality or yield (ill. 96). The general growth of towns during the twentieth century encouraged the sale of rural land around their peripheries, which made way for urban developments. Property within these urban areas became steadily more valuable, as did land within certain proximities, following the location theories of economic geographers of the 1960s.[14]

Urban maps

These increased in number and detail as the supply and market for capital became centred within towns. In every urban expansion of the twentieth century, be it the creation of purpose-built towns such as Gdynia in Poland (1926, an industrial port), new capital cities such as Canberra in Australia (1913) and Brasília in Brazil (1960, both government administration centres), or holiday conurbations such as Benidorm in Spain (developed dramatically for tourism from the 1960s), maps played a part in the creation and management of their commercial activities. The spaces of towns were mapped at a greater scale than rural areas. The Ordnance Survey of Great Britain Town Series, for example, produced updated revisions of nineteenth-century large-scale town maps at the beginning of the century which showed the locations of both urban and commercial premises. Gazetteers and business directories, and from the 1990s GIS (geographic information system) tools, were used by sales and marketing companies to navigate and solicit business from urban areas

Applied algorithms such as the 'Travelling Salesman Problem' streamlined the logistics of business by calculating the shortest routes between two or more points. Some digital applications and intrusive marketing techniques aroused ethical suspicion by the close of the century.[15] Nevertheless, by then GIS was being used comprehensively in government and commercial management in the public and private sectors, and in marketing, transport, business and environmental management (ill. 97). In 2003 the global worth of the GIS industry was predicted to reach between $3.3 billion and $30 billion by 2005.[16]

95 (left)
L. Dudley Stamp. *Great Britain. Land Utilisation*. Ordnance Survey, 1942. Maps 1135.(64)

96 (next page)
Red Deer Land District: Province of Alberta by J. E. Chalifour, 1914. This map was produced by the Canadian Government, and was accompanied by a land registration certificate acknowledging the purchase of the land parcel identifiable on the map.
Maps CC.5.a.463

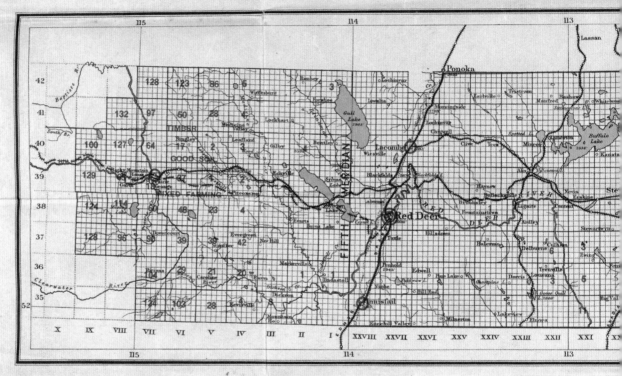

EDITION OF JAN. 1ST, 1914.

NUMBER OF VACANT QUARTER SECTIONS SHOWN
IN RED FIGURES IN EACH TOWNSHIP.

OFFICE OF AGENCY _____ ◎

OFFICES OF SUB-AGENCIES _____ ○

DOMINION LANDS ACT, 1908

All unoccupied surveyed agricultural lands, whether odd-numbered or even-numbered sections, that are not reserved or that have not been disposed of are open to entry for Homestead.

The area enclosed by continuous red line shows lands under which the provisions of

(1) Clause 28, respecting Purchased Homesteads apply; and
(2) Clause 27, Pre-emptions are authorized within such townships as may be designated for that purpose by order of the Governor-in-Council.

RED DEER L

PROVINCE

Scale 1.792.000

J. E. Chalifo

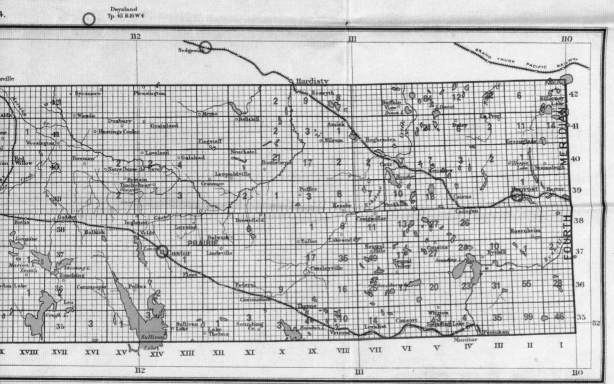

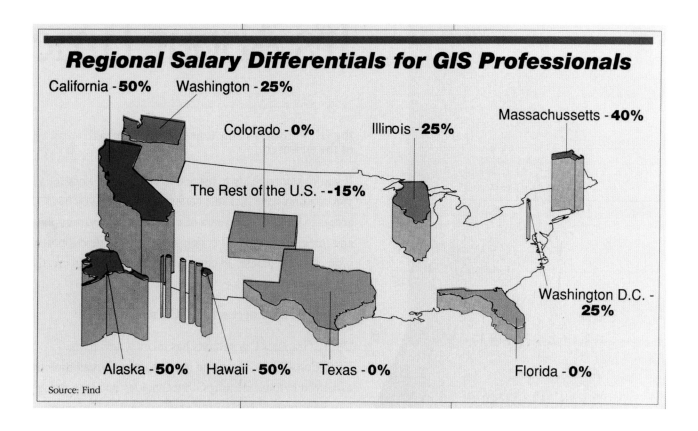

Regional Salary Differentials for GIS Professionals

California - **50%**   Washington - **25%**

Massachussetts - **40%**

Colorado - **0%**   Illinois - **25%**

The Rest of the U.S. - **-15%**

Washington D.C. - **25%**

Alaska - **50%**   Hawaii - **50%**   Texas - **0%**   Florida - **0%**

Source: Find

Fire insurance maps

Made by Charles Goad in Canada, the UK and other parts of the world (from 1875), and by Daniel Alfred Sanborn in the United States (from 1867), these commercial maps showed the external and internal details of town buildings to enable companies to calculate insurance premiums. Interior views included materials and dimensions. Goad went on to specialise in shopping centre maps in Canada and Britain from the 1960s, not coincidentally during a period in which, for the first time since the 1920s, working-class people had spending power as a result of the post-war boom (ill. 98).[17]

Advertising maps

These were among the most effective of the maps employed practically to create profit. Their effectiveness derived from the fact that they were designed to be used by those from whom profit was extracted. The association between maps and advertising was certainly nothing new in 1900. Maps had been used to promote sales at least as early as 1720, when the London map seller George Willdey advertised 'usefull instruments and curiositys ... made to the utmost perfection and sold wholesale or retail at the place above mentioned' in the corner of his printed broadsheets.[18] But in the twentieth century their practical utility increased as maps became more widely used and advertising became more sophisticated. For example, 'Seat of War' maps such as the *Daily Mail*'s of 1914 exhibited perceptive product placement in their inclusion of adverts for alcohol, which dulled the senses during times of combat.

97 (above)
'Regional Salary Differentials for GIS Professionals', *GiS World*, volume 3.6, 1990.
4179.370000

98 (right)
Charles Goad, *Sunderland*, 1968.
Shopping centre plan.
Maps 1175.(224)

SUNDERLAND

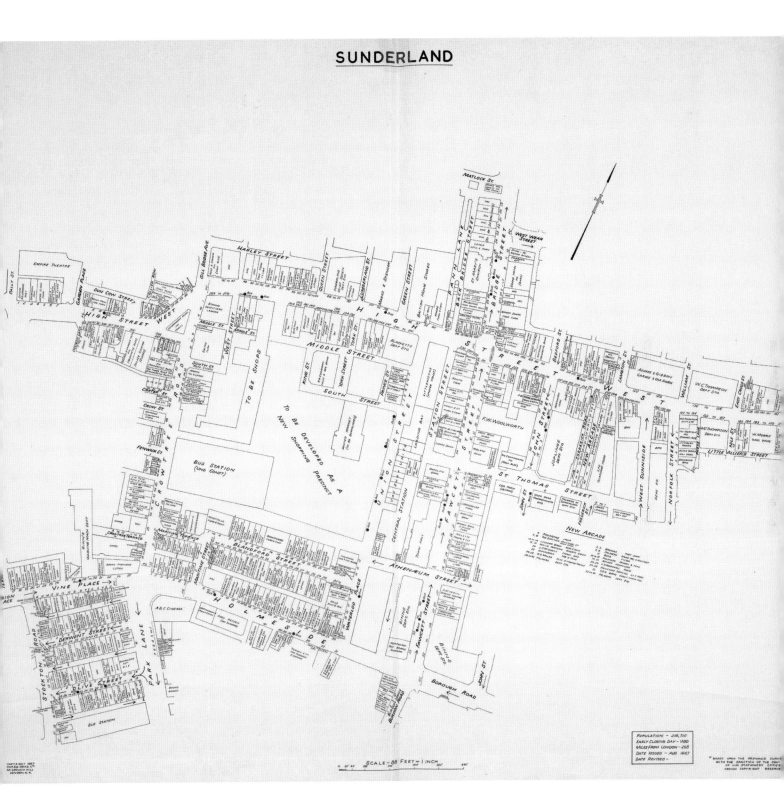

POPULATION – 216,710
EARLY CLOSING DAY – WED
MILES FROM LONDON – 268
DATE ISSUED – AUG 1967
DATE REVISED –

SCALE – 88 FEET = 1 INCH

Lake Texoma on the Texas-Oklahoma Border

99
Texas Road Map. Cover. Esso
Standard Oil Co., 1951.
Maps 197.c.45.(3)

Even more successful was the tactic of giant oil companies such as ESSO, Shell and Standard Oil of printing millions of road maps from the 1890s onwards, and giving them away free as promotional materials for their customers at petrol stations. The road maps were colourful, interesting and useful objects for the increasing numbers of car owners to collect.[19] American maps often included images of happy American families in scenic locations engaging in wholesome family outdoor activities – precisely the scenarios which had motivated the car journey – with the stop-off at the petrol station owned by the oil company advertised on the free map (ill. 99). The maps encouraged happy associations with the product and created a loyal customer base.

Tourist maps

As a reflection of the invigorated tourist industry of the 1950s, tourist maps increased in number. Maps in guidebooks and postcards illustrated commercial air travel routes and holiday destinations, using colour, advertisements and a highly pictorial style to enhance their appeal. The practical function of tourist maps was, as with advertising maps, to facilitate sales. In order to do this, their main device was to create an idealised cartographic image of their chosen location so that it would appeal more powerfully to their customer.[20] In the 1950s a European consumer society wished to escape the experience of the Second World War, and the maps presented just the paradise they craved.[21]

Tourist maps of new destinations such as the Spanish Costa Blanca portrayed seaside resorts as unspoilt and sun-drenched, and this image endured even when the holiday crowds had begun to flock and the weather turned overcast. In Ulster during the 1940s, tourist maps appeared which emphasised the historic and traditional side of Northern Ireland, in stark contrast to the heavily industrial imagery with which it had been promoted before the Second World War (industrial tourism remained popular, especially at the site of Henry Ford's car plant in Illinois; ill. 100). A similar economically motivated image was created by Liverpool City Council, which decided in the early 1970s to reverse the city's flagging fortunes by realigning it with The Beatles, its most successful post-war export (ill. 101).

THE IDEOLOGICAL USE OF MAPS FOR PROFIT

By presenting an idealised version of reality in order to be more commercially effective, tourist maps were using ideological as well as practical methods to create profit. This was nothing new. The ideological use of maps to illustrate and thus support an economic scenario, in turn enabling profit, can be traced back at least as far as early urban maps like Jacopo di' Barbari's 1500 bird's-eye view of Venice. In this map Venice was portrayed in a dramatic and realistic manner, its quaysides bustling with ships and commercial activity. The map communicated the message of Venice's mercantile might to the merchants and courtly audiences who viewed it.[22] Maps validated the existence of economic productivity by illustrating it. The positive ideological message was then applied in the variety of contexts in which maps were used. In the twentieth century these contexts widened considerably from the government office to the school and the domestic setting, enabling maps to support capitalist activity at a variety of levels from the international to the local.

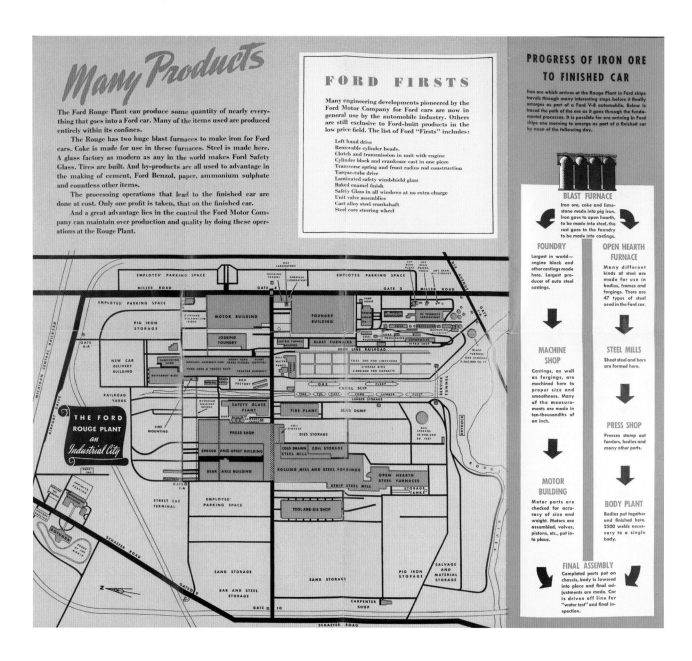

Many Products

The Ford Rouge Plant can produce some quantity of nearly everything that goes into a Ford car. Many of the items used are produced entirely within its confines.

The Rouge has two huge blast furnaces to make iron for Ford cars. Coke is made for use in these furnaces. Steel is made here. A glass factory as modern as any in the world makes Ford Safety Glass. Tires are built. And by-products are all used to advantage in the making of cement, Ford Benzol, paper, ammonium sulphate and countless other items.

The processing operations that lead to the finished car are done at cost. Only one profit is taken, that on the finished car.

And a great advantage lies in the control the Ford Motor Company can maintain over production and quality by doing these operations at the Rouge Plant.

FORD FIRSTS

Many engineering developments pioneered by the Ford Motor Company for Ford cars are now in general use by the automobile industry. Others are still exclusive to Ford-built products in the low price field. The list of Ford "Firsts" includes:

Left hand drive
Removable cylinder heads
Clutch and transmission in unit with engine
Cylinder block and crankcase cast in one piece
Transverse spring and front radius rod construction
Torque-tube drive
Laminated safety windshield glass
Baked enamel finish
Safety Glass in all windows at no extra charge
Unit valve assemblies
Cast alloy steel crankshaft
Steel-core steering wheel

PROGRESS OF IRON ORE TO FINISHED CAR

Iron ore which arrives at the Rouge Plant in Ford ships travels through many interesting steps before it finally emerges as part of a Ford V-8 automobile. Below is traced the path of the ore as it goes through the fundamental processes. It is possible for ore arriving in Ford ships one morning to emerge as part of a finished car by noon of the following day.

BLAST FURNACE
Iron ore, coke and limestone made into pig iron. Iron goes to open hearth, to be made into steel, the rest goes to the foundry to be made into castings.

FOUNDRY
Largest in world—engine block and other castings made here. Largest producer of auto steel castings.

OPEN HEARTH FURNACE
Many different kinds of steel are made for use in bodies, frames and forgings. There are 47 types of steel used in the Ford car.

MACHINE SHOP
Castings, as well as forgings, are machined here to proper size and smoothness. Many of the measurements are made in ten-thousandths of an inch.

STEEL MILLS
Sheet steel and bars are formed here.

PRESS SHOP
Presses stamp out fenders, bodies and many other parts.

MOTOR BUILDING
Motor parts are checked for accuracy of size and weight. Motors are assembled, valves, pistons, etc., put into place.

BODY PLANT
Bodies put together and finished here. 2500 welds necessary to a single body.

FINAL ASSEMBLY
Completed parts put on chassis, body is lowered into place and final adjustments are made. Car is driven off line for "water test" and final inspection.

100 (above)
Souvenir of your Trip Through the Ford Rouge Plant, Ford Motor Co., 1940.
Maps CC.5.b.34.

101 (right)
Beatles map, City of Liverpool Publications Office, 1974.
Maps CC.5.b.53

102 (next page)
MacDonald Gill, *The 'Time & Tide' Map of the Atlantic Charter*.
G. Philip & Son, 1942.
Maps 950.(211)

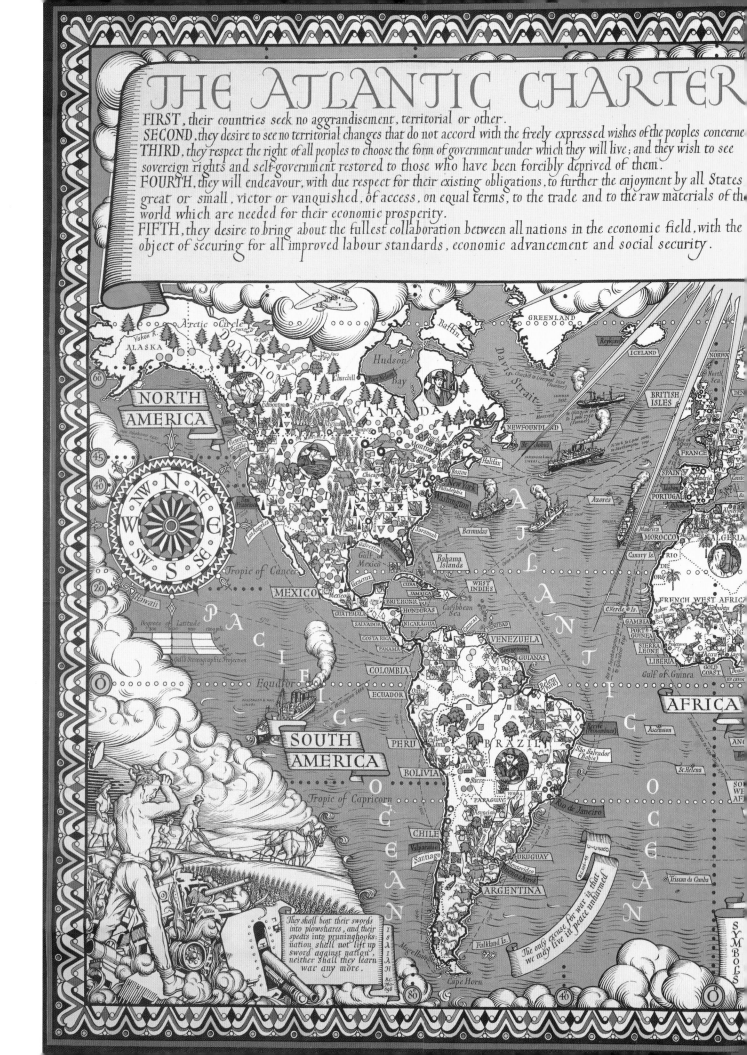

THE ATLANTIC CHARTER

FIRST, their countries seek no aggrandisement, territorial or other.

SECOND, they desire to see no territorial changes that do not accord with the freely expressed wishes of the peoples concerne[d]

THIRD, they respect the right of all peoples to choose the form of government under which they will live; and they wish to see sovereign rights and self-government restored to those who have been forcibly deprived of them.

FOURTH, they will endeavour, with due respect for their existing obligations, to further the enjoyment by all States great or small, victor or vanquished, of access, on equal terms, to the trade and to the raw materials of the world which are needed for their economic prosperity.

FIFTH, they desire to bring about the fullest collaboration between all nations in the economic field, with the object of securing for all improved labour standards, economic advancement and social security.

They shall beat their swords into plowshares, and their spears into pruninghooks: nation shall not lift up sword against nation, neither shall they learn war any more.

ISAIAH B.C. 760-698

The only excuse for war is, that we may live in peace unharmed

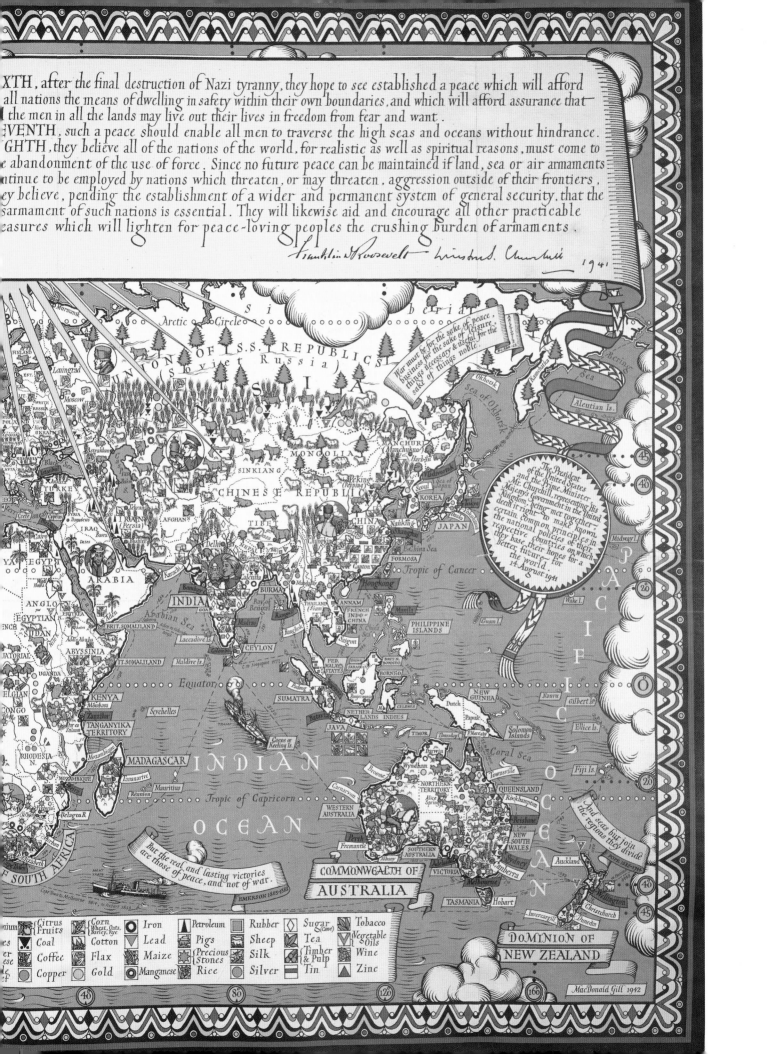

International economic ideology

This was communicated through maps at critical junctures in the twentieth century. These junctures often happened shortly after the end of world conflicts, and the messages implicit in the maps were of economic cooperation between allies. For example, the 1942 *Time and Tide Map* by MacDonald Gill, full of economic illustrations, described the principles of the Allied post-war settlement including 'the fullest collaboration between all nations in the economic field' (ill. 102). Barclays Bank's *The World We Live In* world map of 1958 was accompanied by an essay by the geographer David Linton arguing for international scientific cooperation to cater for a vastly expanding global population.[23] Maps and atlases were commissioned by post-war international bodies such as the World Bank, while the logo of the United Nations, which included a polar world projection emphasising the positive proximity of the world's landmasses to each other, was emblazoned on the numerous reports of its Economic and Social Council, established in 1946.[24]

Colonial economic propaganda

During the twentieth century this was used extensively to promote imperialist economic ideology. For example, posters were used to proclaim the importance of the industrial Don Basin in Ukraine to the Soviet Union during the 1920s, as a means of maintaining patriotism and support.[25] In a similar way, during the Second World War the British Empire Marketing Board commissioned a poster of Australia by MacDonald Gill which celebrated Australia's productivity and importance to the Allied war effort, in order to maintain British as well as Australian economic support and morale. Colourful maps of Spanish colonies such as Equatorial Guinea were produced during the 1940s to promote economic justification for continued Spanish control of its colonies. Economic propaganda could also be used to coerce, as with a European Union map which pressed the benefits of membership upon sceptical populations such as that of the UK in the late 1970s (ill. 103).

National atlases

These provided ideological support to national economies, which multiplied in number as former colonial states asserted their independence from European rule after 1945. The demarcated national borders, which in the case of Africa had often been carelessly drawn by previous colonial administrations, nevertheless identified ownership of land and natural resources, and encouraged disputes, for example between Malawi and Tanzania: the *National Atlas of Malawi* (1983) claimed the disputed land.

National atlases proudly presented economic statements and communicated the advances science and technology had provided for national industry and agriculture. Atlases of Canada (1906, with further editions in 1958, 1974, 1980 and beyond), Indonesia (1960), Britain (1963), Greece (1966) and Ivory Coast (1975) were designed as political statements and high-end status symbols. Struggling economies tended to be portrayed in the same way as successful economies, their positive image intended to engender confidence.

103

Europe in Britain: Grants and Loans from the European Community 1973–76. Commission of the European Communities, c. 1977. Maps X.1476

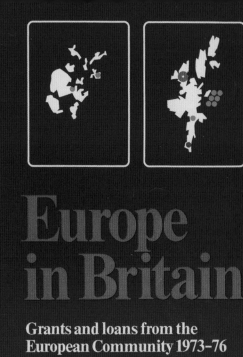

Europe in Britain

Grants and loans from the European Community 1973–76

European Investment Bank		
£561 million	Specific loans	Global loans
Over £5 m		
£1 m to £5 m		
Under £1 m		nil
ECSC/Coal and Steel		
£613 million	Specific loans	Global loans
Over £5 m		
£1 m to £5 m		
Under £1 m		
£34 million	Specific grants	Global grants
Over £5 m		nil
£1 m to £5 m		
Under £1 m		
Agricultural		
£41 million	Specific grants	Global grants
Over £5 m	nil	nil
£1 m to £5 m		
Under £1 m		
Social		
£136 million	Specific grants	Global grants
Over £5 m	nil	
£1 m to £5 m	nil	
Under £1 m		
Regional		
£95 million	Specific grants	
Over £5 m		
£1 m to £5 m		
Under £1 m		
Total amount of Regional grants		
Over £10 m	£5 m to £10 m	To £5 m

Hydrocarbon grants
£6 million

MAPS X.1476.

НАГЛЯДНАЯ КАРТА ЕВРОПЕЙСКОЙ РОССІИ

Составлена
М. И. ТОМАСИКОМЪ.

ДОПОЛНЕНА И ИЗДАНА КРУЖКОМЪ УЧИТЕЛЕЙ
ПОДЪ РЕДАКЦІЕЮ
В. В. УРУСОВА.

1903г.

УСЛОВНЫЕ ЗНАКИ

Historical maps

Used to promote national, political and economic ideologies, historical maps acquired an increasing financial value from the nineteenth century onwards, when the modern trade in antiquarian mapping was established. In 2003 an example of a 1507 world map by the German mapmaker Martin Waldseemüller, the earliest known map to name 'America', was purchased for the Library of Congress for $10 million, which remains the highest sum paid for a single map.[26] The sale bears close comparison with the purchase by Isabella of Spain in 1853 of the Juan de la Cosa chart of 1500 for 4,000 francs – the earliest surviving map to show the New World. Meanwhile the *Mercator Atlas of Europe* of 1570–1 was purchased in 1979 by British Rail as a means of protecting its pension fund against inflation and economic uncertainty. In 1997 the British Library purchased it for well in excess of £500,000.[27] Old, rare and nationally significant maps continued to be valuable commodities, and even investments.

Economic and commercial maps and atlases

These were the most frequently produced map products from the second half of the nineteenth century and throughout the twentieth century. They were used by the full range of business, academic, popular and educational audiences. 'Commercial', 'trade' and 'economic' atlases contained maps with the predominant features of raw materials, agriculture, livestock, manufacture, commodities, shipping routes and telegraph and communication lines. The maps were supplemented by financial graphs, charts and tables. Atlases and maps such as Philips's *Mercantile Marine Atlas* (1904–), Rand McNally's *Commercial Atlas of America* (1926–) and the Horizons de France's *Algérie: Atlas Historique, Géographique et Économique* (1935–) illustrated and thus reaffirmed the economic ideology of the world and its regions.

The frequency of their production demonstrates the prime importance of economic considerations over all others, including environmental and social considerations. It provides a framework for understanding the economic basis for political decisions of the twentieth century. The frequency also gives an insight into the prevailing economic climates of the time. For example, it is less surprising that commercial atlases and maps were produced during the high point of the imperial period (i.e. before 1914) than that their production continued, and even increased, during the world economic crisis and slump of 1929–32. This fact reflects the attempts of world economies to simply 'produce' themselves out of depression.[28]

Educational maps

Instilling a geographical imagination in pupils, these created successive generations of map customers with a sustained appetite for maps. Economic geography was primary, since it was seen to infuse every other type of geographical knowledge. For example, geography education taught the heirs of empire to see in every geological map the coal seams and natural resources to exploit.[29] In wall maps and atlases the use of the Mercator projection emphasised the centrality of Europe; the colouring of the world by imperial possession reflected international trade zones; urban areas and ports identified

104
Наглядная карта Европейской
России (A Pictorial Map of European
Russia), by M.I. Tomasik, 1903.
Maps Roll 537

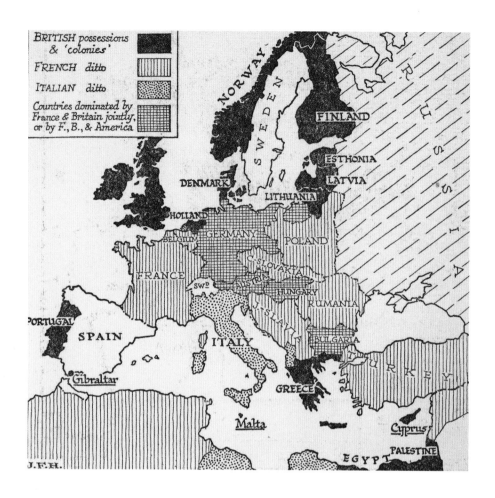

105 (left)
J. F. Horrabin, 'The New Map of Europe', *The Plebs Atlas: Containing 58 Maps for the Use of Students in the Classes of the National Council of Labour Colleges.* Pleb's League, 1926. Maps 47.b.9

106 (right)
Cover of *New Society*, no. 426 (26 November 1970.) 'Fat Cat Britain' illustration by Ralph Steadman. zc.9.d.559

zones of production; and communication routes reflected the flow of goods and capital from colonies to the heartland. Maps such as *Philips' Comparative Wall Atlas of Commercial Development* (1922–) reinforced the geography of economics, and framed the world as a commercial resource in the schoolroom where the vision was most potent for the future administrators of empire.

Educational maps presented economic ideology in different ways for different audiences. For example, the pictorial map of European Russia produced in Warsaw by M. I. Tomasik in 1903 (ill. 104) provided full-blown colour illustrations of Greater Russia's agriculture and industry for the impressionable Russian, Polish or Finnish schoolchild. For a more mature audience, economic maps such as those in Stanley Harrop's *World Wealth in Maps* (1948) used symbols and figures to communicate an efficient and streamlined ideology of business. Meanwhile, the National Council of Labour Colleges (NCLC) and Frank Horrabin's *Plebs Atlas* of 1926 contained simple unadorned black-on-white maps of the world's geography (ill. 105). The reasons for the atlas's simplicity were to save production costs so it could be sold cheaply, and to enable the content of the maps to be effectively communicated.

The ideology of the *Plebs Atlas* was different from that of the majority of its contemporaries.[30] Instead of enabling profit through an illustration of world economic geography, the atlas was concerned with explaining the political context of global

NEWsociety

26 Nov 1970 No. 426 2s (10p) weekly

M.J. Sharpston	THE ECONOMICS OF CORRUPTION
Margaret Morris	HOW FAIR ARE 'FAIR RENTS'?
Reyner Banham	ON THE LEGER EXHIBITION
	SPECIAL FICTION ISSUE
John Gretton	AMERICANS HERE

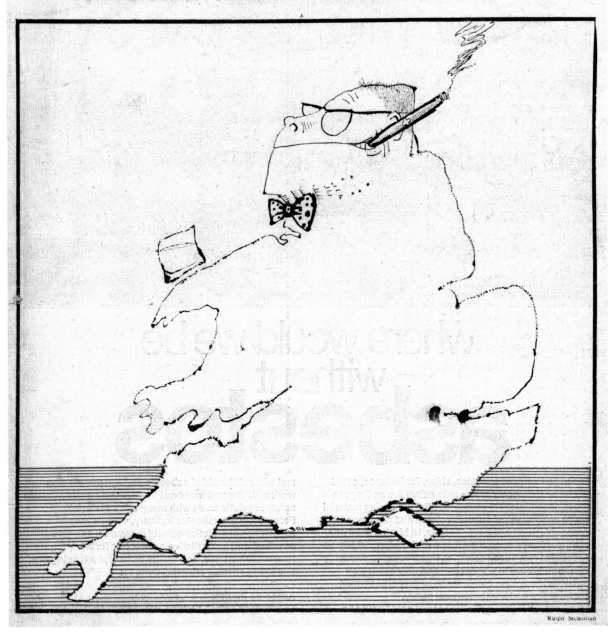

Ralph Steadman

UNIVERSITY OF CALIF., BERKELEY DEFENSE CONTRACTS

COLOR KEY ▫ dept. of defense
▪ atomic energy comm.

TOTAL DEFENSE CONTRACTS
AT U.C. BERKELEY

DOD $9,991,980.00
AEC $1,220,034.00 ON CAMPUS
256,649,000.00 LRL
LIVERMORE
LOS ALAMOS

1. CORY		
Elect. Engr.	DOD	1,092,818.
	AEC	67,243
Elect. Res.	DOD	22,000
2. ETCHEVERRY		
Aero. Engr.	DOD	137,000
Ap. Mech. Engr.	DOD	260,800
Ind. Engr.	DOD	230,175
Mech. Engr.	DOD	368,565
Mech. Engr. Therm.	DOD	86,483
Nuclear Engr.	DOD	26,700
	AEC	30,186
3. LE CONTE		
Physics	DOD	258,318
	AEC	355,102
4. McLAUGHLIN		
Engr.	DOD	42,642
Hy. & Sani. Engr.	DOD	195,368
Struc. Engr.	DOD	204,911
Trans. Engr.	DOD	81,026
5. CAMPBELL		
Astronomy	DOD	13,000
Pure & Ap. Math	DOD	55,596
	AEC	76,000
Computer Res.	DOD	57,260
Mathematics	DOD	48,312
Radio Astro Lab	DOD	76,863
Statistics	DOD	155,129
6. SPACE SCIENCE	DOD	365,293
(off campus)	AEC	57,952
7. LATIMER		
Chemistry	DOD	326,995
8. HILGARD		
Soils & plant nut.	DOD	66,000
	AEC	182,351
9. LIFE SCIENCES		
Botany	AEC	67,000
Phys. & anatomy	AEC	93,621
White Mt. Res.	DOD	31,337
Zoology	DOD	16,395
	AEC	18,298
10. NAVAL ARCHITECT.		
Engr.	DOD	151,744
Coastguard	DOD	7,796
11. TOLMAN	DOD	122,335
12. BARROWS		
Res. Mtg. Sci.	DOD	20,057
Economics	DOD	17,500
Inst. Bus. & Econ.	DOD	79,300
13. MOSES		
Letters & Sci.	AEC	115,612
14. EARTH SCIENCE		
Geo. & Geophys.	DOD	84,904
Seismology	DOD	10,000
	AEC	19,000
15. GIAUQUE		
Chem. Engr.	DOD	108,400
16. MORGAN		
Nutri. Sci.	DOD	68,234
	AEC	8,700
17. BIOCHEMISTRY	AEC	70,162
18. DWINELLE		
Linguistics	DOD	70,000
19. HEARST MINING		
Sci. & Engr. Mater.	DOD	27,460
20. WELLMAN		
Entomology	AEC	27,000
21. MULFORD		
Genetics	AEC	21,000
22. CALLAGHAN		
Naval Sci.	DOD	18,349
23. DONNER LAB(LRL)		
Mech. Phys.	DOD	5,000
Rad. Lab.	AEC	10,807
24. OPTOMETRY	DOD	35,000
25. INST. INTERNAT. STUDIES	DOD	49,559
26. Harmon Gym		
27. NAVAL BIO. LAB	DOD	4,054,71X
(off campus)		
28. UNIVERSITY HALL:		
Committee on spec. projects		
Regents		
Hitch		

capitalism to a worker audience that Marxist theory positioned at the supporting base of a hierarchical social structure. The purpose of the atlas was not to increase profit through ideology, but to explain the economic geography in such as a way as to reveal and thus undermine that ideology.

3. Anti-capitalist mapping

Works such as the *Plebs Atlas* numbered in the minority. The majority of maps continued to enable profit through practical and ideological use up to and beyond the economic downturns of the 1970s. But whereas in the 1920s and 1930s there were only cartoonists and communists to complain about capitalist exploitation and economic slump, by the 1970s many more people in society had maps at their fingertips, were familiar with maps and knew how to make, market and exploit them (ill. 106). By the 1970s the very qualities that had made maps such successful profit-making tools were used to work against what was now coming to be seen by the educated socialist consumer as the corrosive role of capitalism in the world. People with no memory of anything other than prosperity and the welfare state were both able and motivated to make maps as tools of protest.

Protest maps against the perceived failure of capitalism merged with other issues which themselves had economic roots, such as the Vietnam War and the global arms trade. Among them, a 1971 map of the University of California at Berkeley campus in identified research units in the pay of the United States military (ill. 107). Similarly cheap photocopied maps were used by groups such as the Campaign for Nuclear Disarmament (CND) in demonstrations against the use of nuclear weapons.

Nuclear protest intersected a growing environmentalism, born of the late 1960s, which turned the earlier century's obsession with natural resources on its head by demonstrating the harm that industry was doing to the environment. From the 1980s, mapping agencies like the Environmental Systems Research Institute (ESRI) were able to record environmental change such as deforestation in their digital mapping, changes which were also visible in television pictures of oil spills (from the tanker *Exxon Valdez* in 1989, for example) and the fallout of nuclear disasters such as that at Chernobyl in Ukraine (1986).

Of all environmental protest imagery, the first images of the earth taken from space by the US Apollo programme in 1968 proved to be the most potent. *Earthrise* appeared to show the Earth hanging alone and fragile in the blackness of space. It was utilised by a range of protest groups and was exploited, for example, in contemporary circular pin badges bearing slogans such as 'Pollution is our industry' (ill. 108). Environmental protest became the focus of a wide range of causes which used maps to articulate rage against the ills of the capitalist world. Maps were also employed within the gallery setting, with the artist Öyvind Fahlström creating a version of the board game Monopoly in 1974 in which each property space represented a state the USA had either invaded, or with which it had an arms agreement.

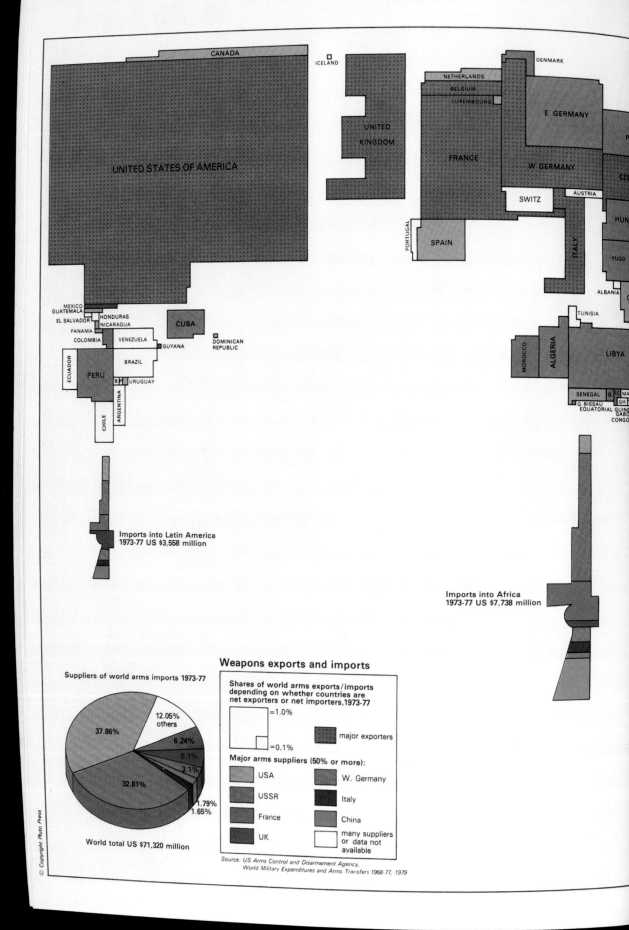

CANADA
ICELAND
DENMARK
NETHERLANDS
BELGIUM
LUXEMBOURG
E GERMANY
UNITED KINGDOM
UNITED STATES OF AMERICA
FRANCE
W. GERMANY
SWITZ
AUSTRIA
CZE
HUN
PORTUGAL
SPAIN
ITALY
YUGO
ALBANIA
TUNISIA
MEXICO
GUATEMALA
HONDURAS
EL SALVADOR
NICARAGUA
CUBA
PANAMA
COLOMBIA
VENEZUELA
GUYANA
DOMINICAN REPUBLIC
MOROCCO
ALGERIA
LIBYA
ECUADOR
BRAZIL
PERU
B
URUGUAY
SENEGAL
G
IC
MA
G·BISSAU
GH
ARGENTINA
EQUATORIAL GUINE
GABO
CHILE
CONGO

Imports into Latin America
1973-77 US $3,558 million

Imports into Africa
1973-77 US $7,738 million

Weapons exports and imports

Suppliers of world arms imports 1973-77

37.86%
12.05% others
6.24%
5.1%
3.1%
32.81%
1.79%
1.65%

World total US $71,320 million

Shares of world arms exports/imports depending on whether countries are net exporters or net importers, 1973-77

☐ = 1.0%
☐ = 0.1%
major exporters

Major arms suppliers (50% or more):

USA	W. Germany
USSR	Italy
France	China
UK	many suppliers or data not available

Source: US Arms Control and Disarmament Agency,
World Military Expenditures and Arms Transfers 1968-77, 1979

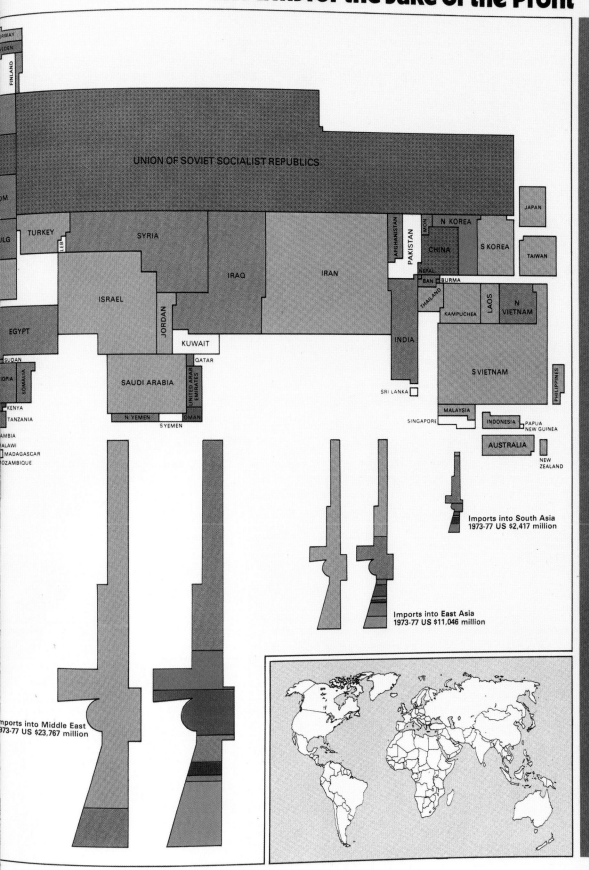

NORWAY

SWEDEN

FINLAND

UNION OF SOVIET SOCIALIST REPUBLICS

...M

BULG

TURKEY

LEB

SYRIA

AFGHANISTAN

PAKISTAN

MON

N KOREA

JAPAN

CHINA

S KOREA

TAIWAN

ISRAEL

IRAQ

IRAN

NEPAL

BAN BURMA

JORDAN

THAILAND

LAOS

N VIETNAM

EGYPT

KUWAIT

KAMPUCHEA

QATAR

INDIA

SUDAN

...IOPIA

SOMALIA

SAUDI ARABIA

UNITED ARAB EMIRATES

S VIETNAM

PHILIPPINES

KENYA

SRI LANKA

TANZANIA

N YEMEN

OMAN

MALAYSIA

...AMBIA

S YEMEN

SINGAPORE

INDONESIA

PAPUA NEW GUINEA

...ALAWI

MADAGASCAR

AUSTRALIA

...OZAMBIQUE

NEW ZEALAND

Imports into South Asia
1973-77 US $2,417 million

Imports into East Asia
1973-77 US $11,046 million

Imports into Middle East
1973-77 US $23,767 million

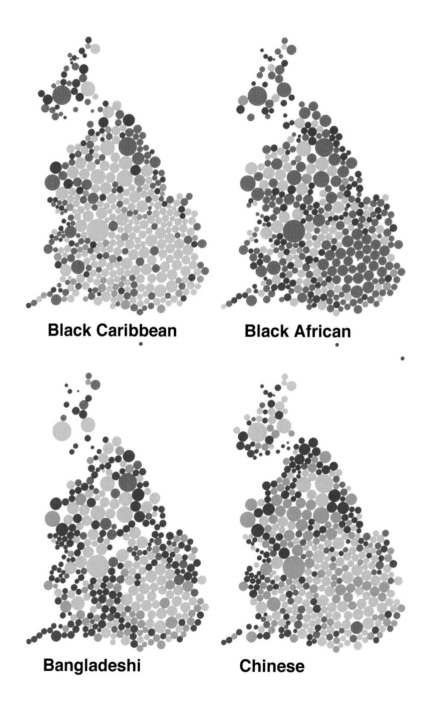

Black Caribbean　　　**Black African**

Bangladeshi　　　**Chinese**

110
Danny Dorling, 'Unemployment by
Ethnic Minority Group', *A New Social
Atlas of Britain*. Wiley, 1995.
British Library Maps 194.b.40

Unemployment by Ethnic Minority Group 1991

proportion of district workforces aged under 25

Scale

☐ = 1 000 000 people

Black Other

Indian

Pakistani

% above the average rate
for 16 to 24 year old
residents out of work

up to -10%

-10% to 0%

0% to 10%

10% to 25%

25% & above

Other Asian

Other Other

Irish born

More conventional maps also bore the backlash. For example, the left-leaning publishing house Pluto Press published *The State of the World Atlas* from 1973 (ill. 109). Subverting the principles of the great commercial, wealth and economic atlases of the earlier century, it presented not world commodities or energy maps but maps of world pollution, offshore tax havens and world arms sales. In a similar vein, the cartographer Arno Peters published a world map in 1972 using a late nineteenth-century map projection which diminished in size the areas of Europe and North America. Peters set it in explicit opposition to Gerhard Mercator's famous sixteenth-century map projection, which had emphasised these areas and so, according to Peters, distorted 'the picture of the world to the advantage of the colonial masters of the time'.[31] It was supposed to be a direct attack upon the apparently crumbling imperial world order.

Digital mapping technologies provided further ability to subvert the established capitalist system. In particular, digitally constructed cartograms were able to produce maps scaled according to human factors such as unemployment or ethnic background. Maps included in the social geographer Danny Dorling's *A New Social Atlas of Britain* of 1995 (ill. 110) highlighted the socio-economic conditions of formerly invisible urban populations where the gap between rich and poor was most pronounced. Poor inner-city areas duly assumed far greater prominence on the map than formerly large swathes of privately owned Devon and Sutherland.[32]

Yet, despite attempts to steer the bias of maps away from the economic elite, the established order retained control of mapping systems and databases. It was discovered that maps produced for altruistic purposes such as environmental management could also be exploited for profit.[33] Peters's projection as well as Mercator's struggled to describe a global economic picture in which distances and straight lines, so crucial in the mercantile and imperial era had become irrelevant. The power and influence of international corporations now transcended national borders. Exclusive economic production and free trade zones, such as Manaus in Brazil, could be constructed in the middle of rainforests because transportation was so cheap.[34] The wealth of the richest was kept hidden on tropical islands, away from states that would demand payment of tax.

Ahead of the game, during the economic crises of the 1970s and early 1980s, international corporations and businesses largely succeeded in moving themselves off the map and remaining there in offshore tax havens or in the middle of cities, protected by security, surveillance and tax loopholes. Also off the map, although in a different place, remained the dispossessed, in unmapped shanty towns or outside state borders. These included Mexican migrant workers in the USA, and southern and eastern European migrant workers in the European Union.[35] Symbolic of such exclusion was how the map included on the EU's first euro coin of 2002 showed a Europe with blank spaces where non-EU member states and North Africa would normally be (ill. 111).

Cartographic erasures have never been more pertinent than in the context of the economic history of the twentieth century. Heavy industries in Europe and North America, some of them long unproductive, closed in the face of depression, economic decline and foreign competition, first during the 1930s and then in the 1970s; yet many

111 (right)

The one Euro coin from 2002,

with its map that excludes non-EU

member states and North Africa

of these areas – such as the North American Rust Belt – were once marked proudly and prominently on economic maps and atlases. The erasures were not simply from the map. Aerial and satellite images taken of the UK around the millennium showed former shipbuilding towns with cosmetically grassed-over river banks, mining towns with contoured and now green spoil heaps, and former industrial towns with derelict or cleared inner-city factory sites.[36]

Occasionally, closed collieries and railway lines reopened as heritage museums, with tourist guide maps that could not but replace the former reality with a sanitised, rose-tinted update; such bland disjuncture was in keeping with the *fin de siècle* malaise.[37] Within this vacuum the majority of maps were able to perpetuate the positive ideology of capitalism with greater conviction in advertising, marketing and statements by politicians promising 'road maps' of economic recovery.[38] As in the twentieth century, cartography continued to attempt to smooth over and iron out the economic peaks and troughs in the new millennium.

5

MOVEMENT
Mapping Mobility to Mobile Mapping

Nick Baron

Modern cartography emerged in the sixteenth to the eighteenth century as a means by which 'enlightened' monarchs and their ministers could know the territories over which they ruled. With knowledge of the spatial expanse, structures and limits of the state, its rulers could demarcate and defend the country's borders, see the distribution of its inhabitants, use the land better and develop new transportation and communication links. Mapping of lands overseas was a prerequisite for colonial expansion, settlement and exploitation. Modern cartography, then, was predicated on and reinforced a rational vision of bounded territories, rooted populations, ordered space and regular time (ill. 113). As geographer David Harvey has written: 'maps, stripped of all elements of fantasy and religious belief, as well as any sign of the experiences involved in their production, had become abstract and strictly functional systems for the factual ordering of phenomena in space.'[1]

Yet by the mid-nineteenth century the world was changing. Industrialisation, urbanisation and other modernising processes were dissolving absolutes and certainties, creating a new sense of a dynamic present and prompting new dreams for the future. As Karl Marx and Friedrich Engels – perspicacious observers of contemporary reality – noted in *The Communist Manifesto* (1848): 'All fixed, fast-frozen relations, with their train of ancient and venerable prejudices and opinions, are swept away, all new-formed ones become antiquated before they can ossify. All that is solid melts into air.'[2]

In the twentieth century, two world wars, revolutions and the disintegration of empires brought about vast territorial upheavals and displacements of populations. Scientific, technological and economic innovation made possible and generated massively increased flows of commodities, capital and information. The invention and spread of new means of transport and telecommunications accelerated time and compressed space.[3] Change in the twentieth century occurred at a scale and with an intensifying rapidity that preceding generations could barely have imagined. Political systems, societies, economies and cultures globally sought to come to terms with these fundamental transformations in many, often contradictory, ways – accommodating

112
William Ander, *Earthrise*.
Photograph, 1968.
NASA, AS8-14-2383

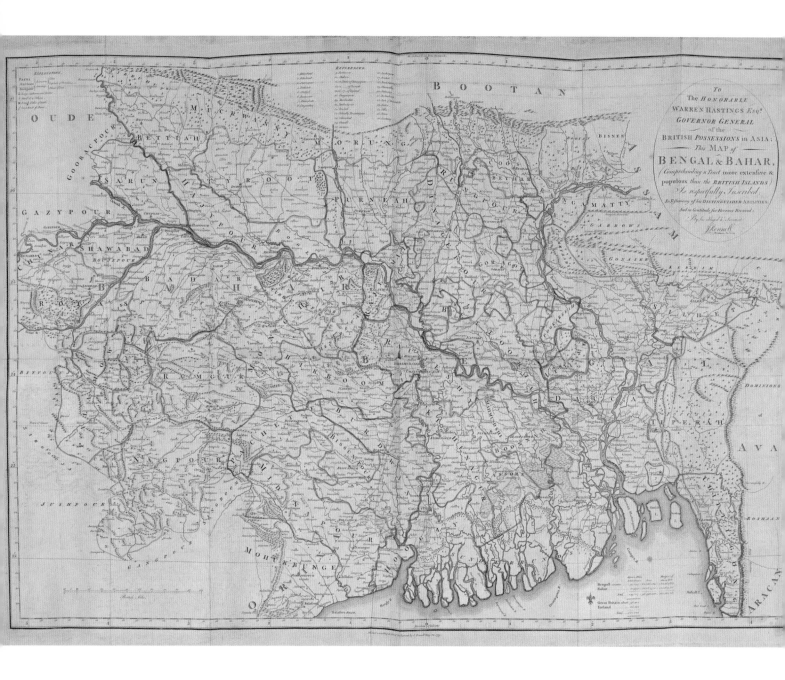

113

James Rennell, 'An Actual Survey of
the Provinces of Bengal, Behar &c.',
A Bengal Atlas..., 1781.
Maps C.25.b.8

to change or striving to plan and regulate development; promoting the movement of money, commerce, people and ideas or attempting to curb mobility and migration; establishing transnational regimes of governance and law or reasserting and reinforcing the nation-state.

Governments in the twentieth century sought both to harness the new possibilities for their own advantage and to defend themselves against the disruptive potential and power of the new. In the words of late twentieth-century cultural theorists Gilles Deleuze and Félix Guattari: '[States still have] a need for fixed paths in well-defined directions, which restrict speed, regulate circulation, relativize movement, and measure in detail the relative movements of subjects and objects.'[4]

This chapter considers how cartography in the twentieth century responded to this new era of flux and flow and reflected the conflicts and tensions that the new age engendered. It is a story of struggle and creativity. On the one hand, mapmakers strove to move beyond cartography's conventionally static framing of ordered space – the map as a snapshot, frozen in time. Throughout the century, they developed or appropriated new technologies, visual forms and aesthetic strategies to capture dynamism and change, thus permitting the sponsors and users of maps better to perceive, comprehend and reap the benefits of continuous innovation. Yet, at the same time, most twentieth-century cartography, in the service of particular ideological, political or commercial interests, endeavoured to create new systems of 'factual ordering' to impart stability and sense to a reality that, for many citizens of the world, now seemed terrifyingly elusive, shifting and transient.

The first section of this chapter considers cartography designed to facilitate movement. Many of these maps in fact served also to constrain or channel mobility. The second section examines a selection of maps created to illustrate motion and change, both in society and in the natural world. These were static maps attempting to depict movement.

The third section looks at maps that were produced dynamically, or that were themselves in motion. During the early and mid-twentieth century, cartographers exploited the latest innovations in transport and communications technologies to establish new means of imaging the Earth's surface. Equipping aeroplanes and later satellites with sophisticated photographic apparatus and mechanisms for the remote transmission of data, cartographers surveyed far greater expanses of territory far more quickly than had previously been possible. Later in the century, travel beyond the Earth's atmosphere enabled humans to produce images of the whole planet, catalysing a new sense of global interconnectedness and a new appreciation of the beauty and fragility of our shared home (ill. 112).[5]

At the same time, the new medium of cinema provided the opportunity for experimentation with techniques for visually capturing and communicating the constant motion of the modern world – for example by filming landscapes from a moving train or aeroplane, or by recording the movement of people in and through a single space over a period of time. Film also introduced and popularised the use of animated maps to narrate temporal shifts and spatial change. Moving pictures brought maps alive.

In the last decades of the twentieth century cartographers, often inspired by cinematic techniques, seized on rapid advances in digital technology as well as in information science to create dynamic visualisations of space that could also be interactive, individualised and user-driven. These were mobile and mutable maps which rendered and conveyed movement in real time, as well as recording the tracks and traces of past itineraries.

In the first decades of the twenty-first century, dynamic digital cartography has become so ubiquitous and assimilated into our everyday lives – for example, through satellite navigation systems, computer games, GPS-enabled smartphones and Google Maps – that traditional paper maps have for many become obsolete, and mapping (as well as being mapped) has become taken for granted, instinctual and often invisible. Practices of mapping – the sensory experiences of observing and traversing space – have become reincorporated into the map. This is a development that is full of promise yet fraught with risk.

Cartography is a social phenomenon, whether it is conceived as science, technology, art form or cultural artefact. As such, it can only be understood in relation to the social structures in which it is embedded and the social spaces which it represents and, indeed, which it constructs through its representations. Think of how the familiar London Underground map (ills 3, 4) or the satellite picture of the River Thames used in the *EastEnders* credits (see Introduction) contribute to forming our 'image' of London. Cartography is powerful. It has the potential to liberate its users, but also to subject them to surveillance and oppression; it may help people and communities to find their own 'place' or 'path' in the world, but it may also serve to dis-locate them. It is up to us to shape the maps of the world in which we wish to live.

Maps as instruments of mobility

Since its earliest days cartography has performed two principal functions: to store and communicate spatial information; and to aid people to navigate on land and sea. During the late nineteenth and early twentieth centuries, as new means of transport – first the railway, later the automobile and then aviation – enabled people to travel ever faster and ever further, so maps as instruments of wayfaring played an ever greater role in the lives of an increasingly wider public. Cartography became an essential and inseparable part of popular culture.

In Britain in the twentieth century, the expansion of car ownership, in particular, generated mass demand for legible and attractive maps and atlases. From 1912 the Automobile Association offered personalised itineraries for its motoring members (see Introduction). In 1934 alone the organisation issued 700,000 of these route maps.[6] These comprised strip maps of the road from A to B, with detailed textual instructions. Occasional comments such as 'Fine valley scenery in places' accompanied the itineraries, but their focus on the road generally eschewed distraction or embellishment. In a similar way, from 1936 the *Geographers' London A–Z Street Atlas,* designed by Phyllis Pearsall

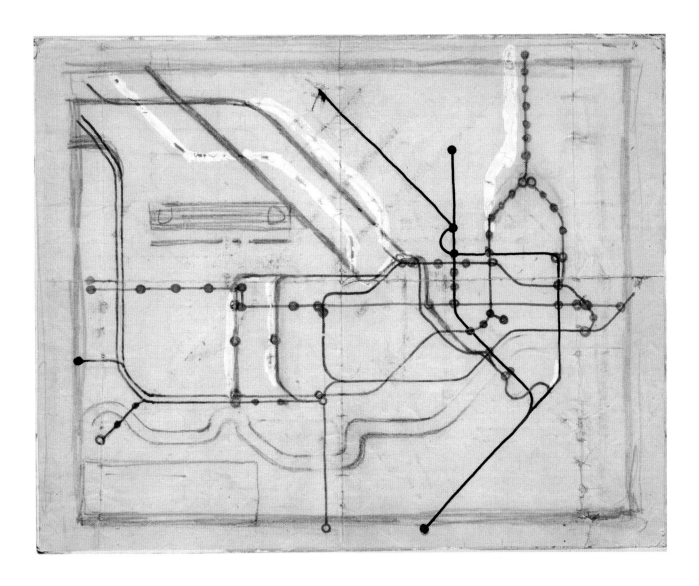

114
Harry Beck, early sketch for the
map of the London underground
railway system, 1931.
Victoria & Albert Museum,
E.814-1979

(1906–96), presented a 'clear and up-to-date' vision of London in its maps, which
excluded all but roads, railways and principal buildings, the spaces in between left blank.[7]
In both cases, the visualisation of space was subordinated to the role of these maps as
instruments of navigation. They were not maps designed to help travellers choose what
to see, to encourage them to venture off the beaten path or to help them apprehend
their environment – only how to navigate their immediate surroundings most rationally
and with the greatest economy.

When London Transport commissioned an employee, Harry Beck (1902–74), to
create a new map of the London Underground railway system in 1931, he produced an
even more abstracted vision of space (ills 114, 115). The design problem he faced was
how to represent the many new stations that had recently been built in the centre of
the network as well as the line extensions, which by then were reaching far out into the
suburbs. Beck, an engineering draftsman, drew on the conventions of electrical circuit
diagrams to resolve this issue. By reducing the map to its fundamental geometries, using
verticals, horizontals and diagonals, and greatly exaggerating the scale of the central area,

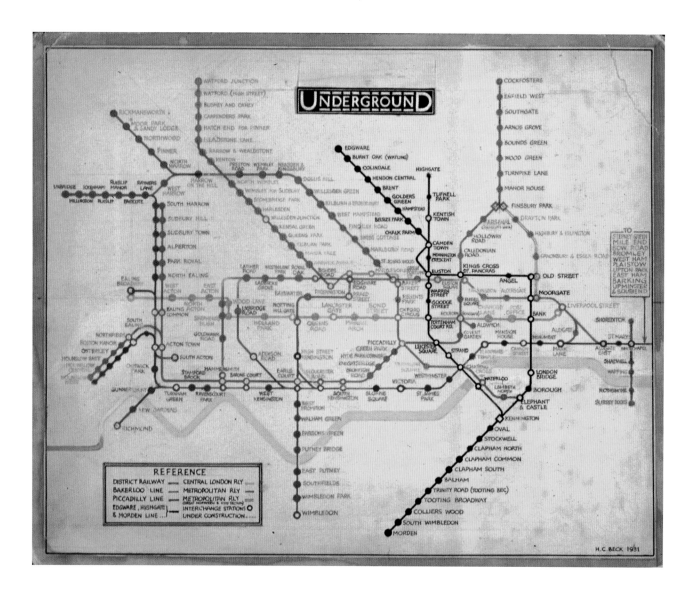

115 (above)

Harry Beck, London Underground map, 1931

Beck enabled it to function efficiently: to be legible, comprehensible and memorable at a glance. Apart from the inclusion of the River Thames, and the station names (which do not appear in the preparatory sketch), the map provided little sense of London's spatial scale or the inter-relations of points on the map, and even less of the city's physical topography, economic activities or social differentiation.[8]

These functional maps construct a highly selective, purposeful rendering of space. They guide travellers, but also regulate and channel their mobility. The limitation of such maps is illustrated by reference to the 'Knowledge' test, which requires prospective London taxi drivers to learn 320 set routes (called 'runs') across the city, within a six-mile radius of Charing Cross, and how to connect tens of thousands of landmarks and points of interest with these runs (ill. 116). Trainee cabbies criss-cross the capital on mopeds, with *London A–Z* map sheets on handlebar-mounted clipboards, gaining an understanding of how the abstract maps relate to the real spaces and rhythms of the city. Acquiring the Knowledge, which can take up to five years, is about combining formal cartographic learning with accumulated experience of travelling. The Knowledge brings the *London*

116 (right)
London trainee cab driver
studying for the 'Knowledge'
test in the 1990s

A–Z maps to life in a way that no satnav can. Indeed, recent scientific research has shown that acquiring the Knowledge – the living map of London – causes taxi drivers' brains to change, physically enhancing both the structures that manage memory storage and those that enable the purposeful recall of spatial images for navigation.[9]

The expansion of mass tourism, at home and abroad, also popularised cartography and made maps more widely accessible. A burgeoning tourist industry was one of the main sponsors of the creation and dissemination of maps of the destinations they were marketing, and a plethora of publications – glossy travel agency brochures, travel guides, atlases, gazetteers and so on – incorporated maps designed not only to help tourists find their way in unfamiliar locations, but also to entice potential travellers by illustrating the attractions of a particular country, region or place; to help them plan their journeys in advance; and to act as aides-memoires for reminiscences about their trip.

The Austrian artist Heinrich Berann (1915–99) played a key role in the evolution of twentieth-century popular cartography. From the mid-1930s, Berann produced hundreds of panoramas that combined the techniques of landscape painting with those of map-drawing.[10] His early views of Alpine tourist resorts won him a commission from the organising committee of the 1956 Winter Olympics in Cortina, northern Italy, to design the event's official map. He subsequently designed panoramic maps for five more Winter Olympics, as well as for the US National Geographic Society and US National Parks (his ocean floor map is discussed below). Berann's poster for the 1976 Winter Olympic Games, held in Innsbruck near his home, exemplifies his technique (ill. 117). Painted with a limited yet vivid palette of colours, this offers a bird's-eye perspective on an idealised view of the town and its surrounding mountains, derived from a careful study of topographical maps and aerial photographs. A vortex of clouds and ethereal light immerses the observer in the picture, lending depth and breadth to the two-dimensional

XII. Olympische Winterspiele Innsbruck

4.–15. Februar 1976

Innsbruck
1 Olympia-Eisstadion mit Schnellaufbahn
2 Kunsteishallen „Messegelände"
3 Bergisel-Spezialsprungschanze
4 Olympisches Dorf

Igls
5 Abfahrtslauf Herren
6 Olympia-Bob- und Rodel-Kunsteisbahn

Axamer Lizum
7 Riesenslalom Herren
8 Slalom Herren
9 Slalom Damen
10 Abfahrtslauf Damen
11 Riesenslalom Damen
12 Abfahrtslauf Herren, Reservestrecke

Seefeld - Telfs
13 Langlaufgebiet und Biathlon
14 Toni-Seelos-Sprungschanze

XIIes Jeux Olympiques d'Hiver Innsbruck

du 4 au 15 février 1976

Innsbruck
1 Stade olympique de patinage avec piste pour les concours de vitesse
2 Patinoire olympique couverte „Messegelände"
3 Tremplin spécial sur le Bergisel
4 Village olympique

Igls
5 Descente hommes
6 Piste olympique de glace artificielle pour bobsleigh et luge

Axamer Lizum
7 Slalom géant hommes
8 Slalom hommes
9 Slalom dames
10 Descente dames
11 Slalom géant dames
12 Descente hommes - Piste de remplacement

Seefeld - Telfs
13 Pistes de fond et du Biathlon
14 Tremplin Toni Seelos

XII Olympic Winter Games Innsbruck

4th–15th February 1976

Innsbruck
1 Olympic ice stadium with speed skating oval
2 Artificial indoor skating rinks „Messegelände"
3 Bergisel special ski jump
4 Olympic village

Igls
5 Men's downhill
6 Artificial olympic bobsleigh and toboggan run

Axamer Lizum
7 Men's giant slalom
8 Men's slalom
9 Ladies' slalom
10 Ladies' downhill
11 Ladies' giant slalom
12 Men's downhill, alternate course

Seefeld - Telfs
13 Cross-country courses and Biathlon
14 Toni Seelos ski jump

XII Зимние Олимпийские Игры Инсбрук

4–15 февраля 1976 года

ИНСБРУК
1 Олимпийский стадион с дорожкой для скоростного бега
2 Искусственные катки в "Мессехалле"
3 Лыжный трамплин в Бергизеле
4 Олимпийская деревня

ИГЛЬС
5 Скоростной спуск для мужчин
6 Искусственная трасса для бобслея и спуска на санках

АКСАМЕР-ЛИЦУМ
7 Гигантский слалом для мужчин
8 Слалом для мужчин
9 Слалом для женщин
10 Скоростной спуск для женщин
11 Гигантский слалом для женщин
12 Скоростной спуск для мужчин, запасная трасса

ЗЕЕФЕЛЬД-ТЕЛЬФС
13 Лыжная трасса для бега на дистанцию и Биатлона
14 Лыжный трамплин имени Тони Зеелоса

Seilbahnen
Téléphérique
Cable Railwa
КАНАТНЫЕ Д
ЛЫЖНЫЕ ЛИ

A 1 Hungerburg
A 2 Nordkettenb
B Glungezerba
C Patscherkof
D Stubaier Gle
E Muttereralm
F Birgitzköpfl
G Hoadl Stand
H 1 Seilbahnen
H 2 Gschwandtk
15 Pressezentru

Tirol Austria

XII. Olympische Winterspiele Innsbruck 1976

Published by Organisationskomitee der XII. Olympischen Winterspiele 1976, Innsbruck – Panorama: H. Berann, Innsbruck-Lans – Gedruckt in Österreich – Imprimé en Autriche – Printed in Austria by F. Sochor, Zell am See.

1060(4.)

117 (left)
Heinrich Berann, XII. *Olympische Winterspiele Innsbruck 1976*, Innsbruck, 1976.
Maps 1060.(4)

118 (next page)
Royal Wedding Souvenir Map of London, Heraldic Heritage Ltd.
Mapledon Press, 1981.
Maps CC.5.b.36

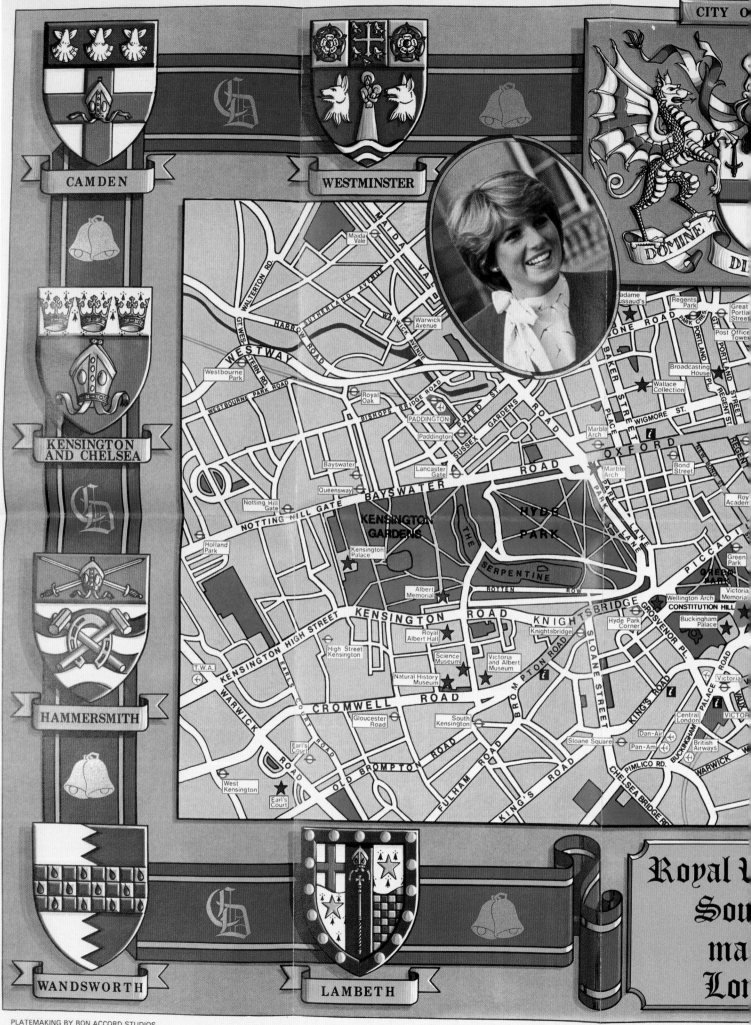

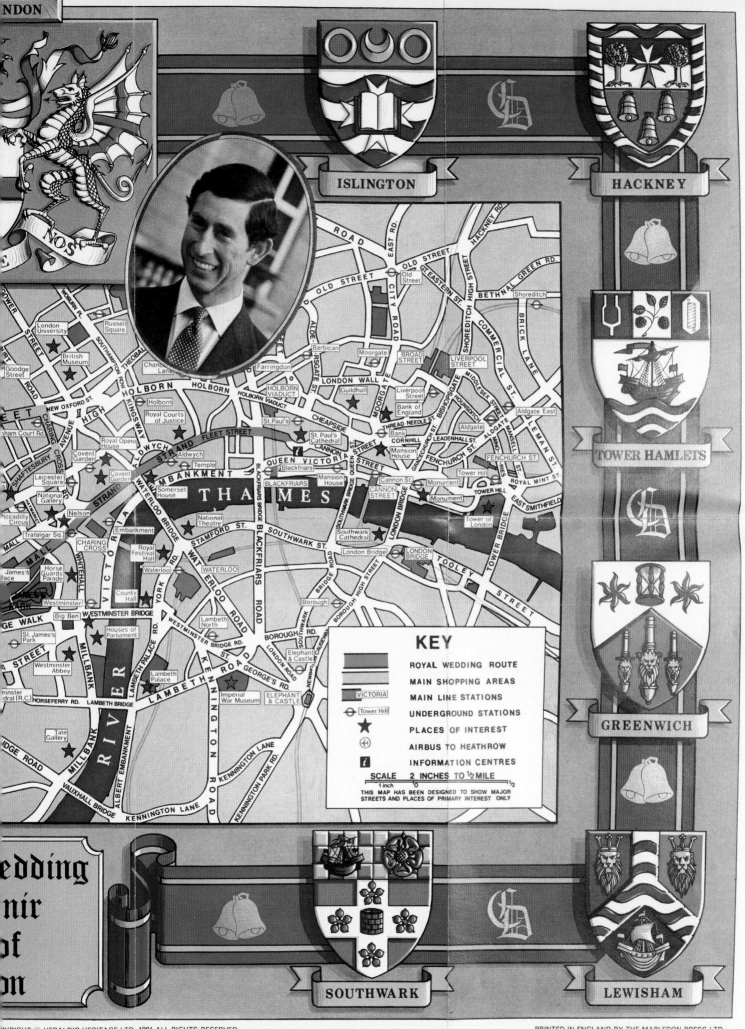

ISLINGTON

HACKNEY

TOWER HAMLETS

GREENWICH

SOUTHWARK

LEWISHAM

THAMES

RIVER

KEY

	ROYAL WEDDING ROUTE
	MAIN SHOPPING AREAS
VICTORIA	MAIN LINE STATIONS
⊖ Tower Hill	UNDERGROUND STATIONS
★	PLACES OF INTEREST
⊕	AIRBUS TO HEATHROW
i	INFORMATION CENTRES

SCALE 2 INCHES TO ½ MILE

1 inch   0   ½

THIS MAP HAS BEEN DESIGNED TO SHOW MAJOR
STREETS AND PLACES OF PRIMARY INTEREST ONLY

image, as well as imparting an otherworldly quality to the landscape. Evidently, the map is only secondarily designed to convey information; principally, it inspires the imagination and allures the traveller.

Berann's panorama achieves its main purpose through its beauty. The 1981 *Royal Wedding Souvenir Map of London* uses a very different style to perform similar functions (ill. 118). This map creates a specific narrative of London's space. It 'sells' London to visitors as a city of heritage and history – the map is framed by the heraldic emblems of the metropolitan boroughs, with the larger crest of the City of London and the portraits of Charles and Diana intruding on and becoming part of the map itself. Prominent green spaces and the blue of the River Thames and the Serpentine soften and 'naturalise' the abstract representation of streets and landmarks. Blue also links these natural features to the Royal Wedding route and to Diana's outfit. At the same time, the map also prompts tourists to 'buy' London, by highlighting the main shopping areas.

The maps and plans of exotic locations published from 1973 within the pages of Tony and Maureen Wheeler's *Lonely Planet* guides were tools for independent travellers finding their way in unfamiliar surroundings. Embedded within textual descriptions of places, the maps, drawn by various authors and illustrators, corresponded closely with the detailed practical information on hotel locations and sites of interest that made *Lonely Planet* the most popular of the various tourist guides of the late twentieth century. The maps were perceived as highly practical, so much so that guidebooks were often confiscated at South Asian airports by authorities nervous about what subversive activities could be performed with such information.

Like the Royal Wedding souvenir map, however, *Lonely Planet* maps were about leisure, consumption and the commodification of spaces associated with tourism. For example, the maps by Geoff Crowther and Linda Fairbairn, particularly the latter's Alice Springs map (ill. 119), included attractive miniature vignettes of key tourist sites drawn in a feathery, pictorial style.

Such tourist maps were designed to stimulate the imagination of armchair travellers as much as to guide wayfarers. Maps have also been frequently used to guide the mind's eye through landscapes that are themselves imaginary. One of the best examples of a work of literature that uses maps to visualise fictional geographies is *The Lord of the Rings*, published 1954 to 1955 by J. R. R. Tolkien (1892–1973). The delicately drawn maps (by the author and his son Christopher) of Middle-earth help the reader follow the complex narrative as it moves back and forth across the land (ill. 120). Maps also guided the author as he constructed his narrative. Tolkien stated that he 'wisely started with a map, and made the story fit'.[11] In this hand-drawn sketch map, a version of which was later published in the final instalment of *The Lord of the Rings*, Tolkien depicts the setting for the climax of the story when the journeys of a number of protagonists come together. By using graph paper he was able to calculate the time needed for them to arrive at their positions and to arrange the story accordingly.

In this section, we have seen how maps designed for route-finding enable and facilitate movement, but also constrain and direct mobility; how they function as instruments of navigation but also as inducements to consumption and stimuli to the

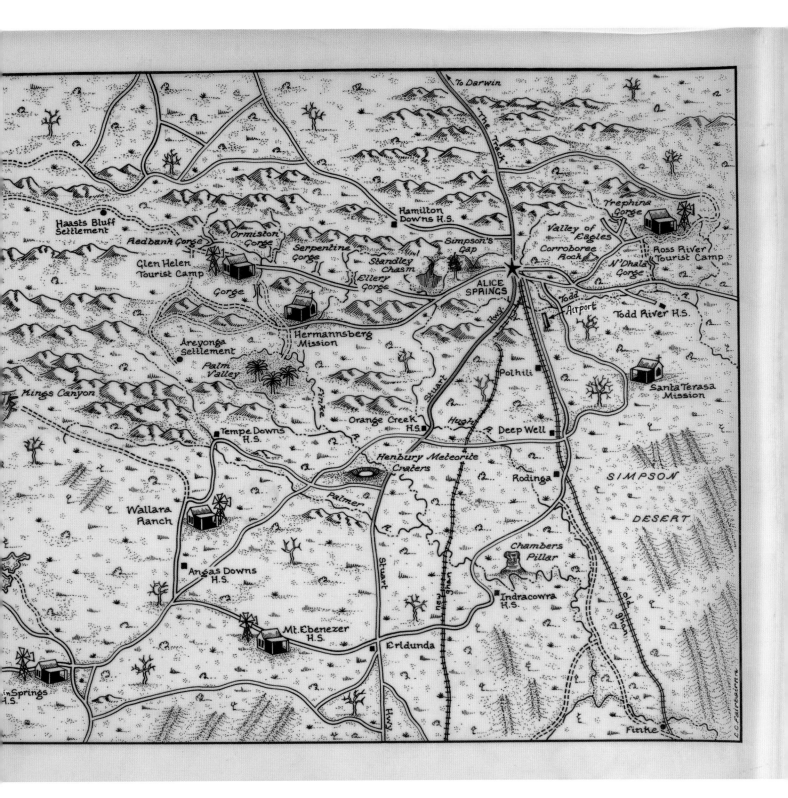

119
Linda Fairburn, 'Alice Springs'. Sketch
reproduced in *Australia: A Travel
Survival Kit*. Lonely Planet, 1983.
Private collection

imagination; and how, in various ways, they construct meaning and perform multiple purposes – some overt, some implicit or clandestine.

We conclude with two maps that illustrate these various functions of wayfinding maps. Both maps purport to represent routes to safety and refuge; in different ways, both maps also disguise danger.

The *Three Mile Island Nuclear Station Evacuation Plan Map* was published within an informational brochure issued by the Pennsylvania Emergency Management Agency in November 1981 (ill. 121). Its production followed an incident at Three Mile Island in March 1979 during which a partial meltdown of a nuclear core released radioactive material into the atmosphere. Numerous inquiries into the event, including a President's Commission, found fault with the official handling of the incident, especially citing poor communications with the population. Roughly 150,000 people voluntarily evacuated the area following the accident.[12] The 1981 map was designed as much to assuage fears and instil calm as to help residents in an emergency. Placed in the glove box of every car, the map reassured readers that there would be no repeat of the confused and panicked movement that had occurred previously. Every element of the map served to soothe: the carefully designated evacuation routes (coloured pink); the identification of 'traffic flow consolidation points' at junctions; and the accompanying text explaining that the 'worst estimated exposure [to radiation] received by an individual during the TMI-2 incident in 1979' was not drastically worse than radiation doses due to natural or other everyday sources. Through the map, the Emergency Management Agency proclaimed its expertise, and told the citizen: 'Trust us!'

In Germany in 1946, millions of citizens were still desperately looking for loved ones who had been uprooted and dispersed in the chaos of the recent world war. For fifty Reichspfennig they could purchase the *Map of Refugee Search Points, Zones of Occupation and Postal Regions* (ill. 122). This leaflet provided them with the current addresses of the offices that administered searches and outlined the procedure for undertaking a search. The map also illustrated the four Allied zones of occupation and the four sectors of Berlin under Allied control, a stark reminder of the country's defeat and division as well as an indication, for the politically aware reader, of emerging Cold War tensions between East and West. Printed below the list of addresses of Refugee and Missing Persons Points is a directive of the Soviet Military Administration in Occupied Germany that centralises under the authority of its Berlin office all search functions within the Soviet zone and city sector, and dissolves all other search agencies, previously administered by provincial authorities and churches. Ostensibly a map produced for humanitarian purposes, it also emphasised Germany's subjugation and the power of the occupying forces, and presaged the suppression of independent civil organisations by the East German communists, with Soviet backing, in subsequent years.

120
Christopher Tolkien's 1948 sketch map of part of Middle-earth
(Rohan, Gondor and Mordor), a version of which was published
in the third book of *The Lord of the Rings* in 1955.
Bodleian Libraries, Oxford, MS. Tolkien S 11/2, fol. 23

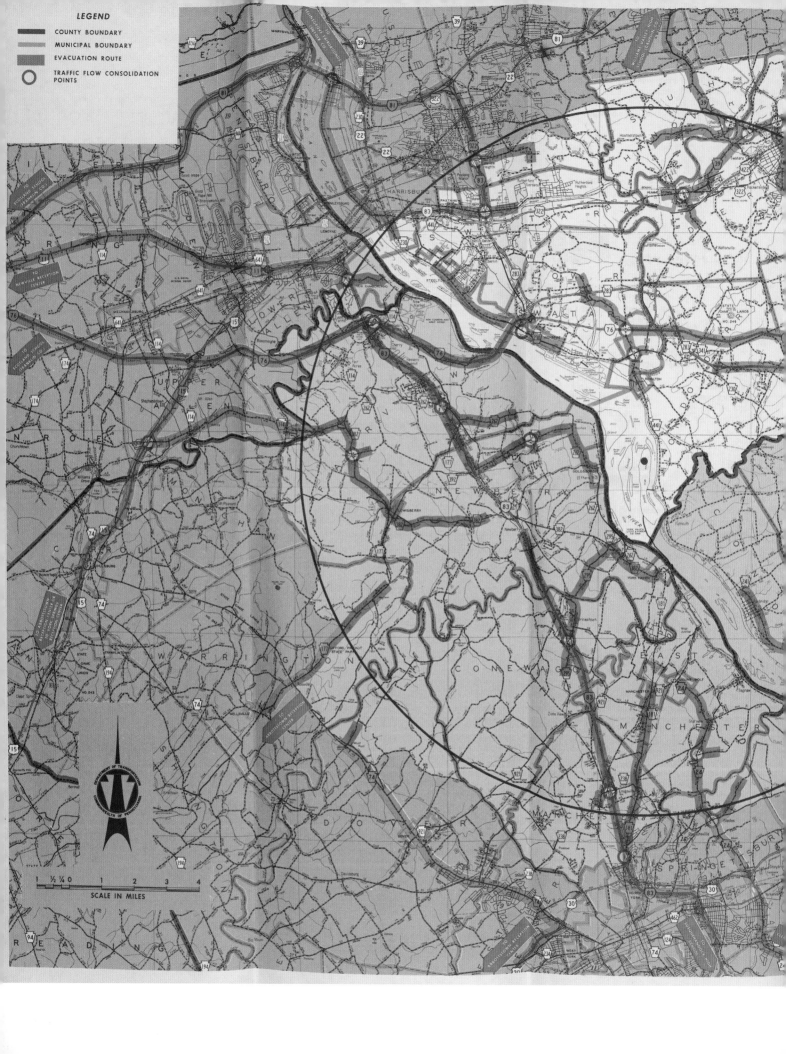

SCALE IN MILES
½ ¼ 0   1   2   3   4

THREE MILE ISLAND NUCLEAR STATION
EVACUATION PLAN MAP
PENNSYLVANIA EMERGENCY MANAGEMENT AGENCY

NOVEMBER 1981

What is Radiation?

Nuclear radiation consists of energy in the form of invisible particles or rays given off by radioactive material. Small amounts of radioactive material occur naturally and always have been part of man's environment. Radioactive materials in varying amounts are present in the earth's crust, the sun's rays, the air we breathe, the food we eat and the water we drink. As a result, every person has radioactive materials within his body. Larger amounts of radioactive materials are produced by and contained within a nuclear power plant.

Man's use of radioactive materials also results in radiation exposure. For example, doctors and scientists have utilized X-rays in medical treatment for many years.

The amount of radiation a person receives is measured in terms of radiation dose. The unit used to measure this dose is called a *millirem*.

The following table shows examples of typical radiation doses due to natural radioactive materials or man's use of radioactive materials compared to the worst estimated exposure received by an individual during the TMI-2 incident in 1979.

	Source	Millirem Per Year
*	Color television	1
*	Airline travel (typical airline passenger who makes 10 flights per year)	3
*	Natural radioactive materials within the body	20
*	Medical X-rays (average patient)	20
*	Cosmic rays	27
*	Natural radioactive materials in the earth	46
**	Maximum offsite exposure during TMI incident	70

* "The Effects on Populations of Exposure to Low Levels of Ionizing Radiation," National Academy of Science, 1980.

** Report of the President's Commission on the accident at Three Mile Island, October 1979, Page **34.**

How Are Incidents Classified?

Should an incident occur at the Three Mile Island Nuclear Station, there are four incident classifications you might hear reported on radio, TV or read in the newspapers. So that you will understand their meaning, they are explained in the order of their potential seriousness:

Unusual Event — Unusual events are in process or have occurred which indicate a potential degradation of the level of safety of the plant. No releases of radioactive material requiring offsite response or monitoring are expected unless further degradation of safety systems occurs.

Alert — Events are in process or have occurred which involve an actual or potential substantial degradation of the level of safety of the plant. Any releases are expected to be limited to small fractions of the Protective Action Guideline exposure levels established by the Federal Environmental Protection Agency (EPA).

Site Emergency — Events are in process or have occurred which involve actual or likely major failures of plant functions needed for protection of the public. Any releases are not expected to exceed EPA Protective Action Guideline exposure levels except near the plant boundary.

General Emergency — Events are in process or have occurred which involve actual or imminent substantial core degradation or melting with potential for loss of containment integrity. Releases can be reasonably expected to exceed EPA Protective Action Guideline exposure levels offsite for more than the immediate plant area.

Karte der deutschen **Flüchtlingssuchstellen, Besatzungszonen** und **Postleitgebiete**

Anschriften der deutschen Flüchtlings- und Vermißten-Suchstellen
sowie die Bestimmungen über den Postverkehr mit Kriegsgefangenen

A. Russische Zone

1) Gesamte Zone	a)	Anschriften Suchdienst der Deutschen Post [10 b] Leipzig C 1
	b)	Suchdienst für vermißte Deutsche in der sowjetischen Okkupationszone Deutschlands [1] Berlin W 8 Kanonierstr. 35
2) Berlin	a)	Hilfswerk der Evangelischen Kirche in Berlin Hauptbüro [1] Berlin W 15 Lietzenburger Straße 36
	b)	Suchdienst des Deutschen Caritasverbandes [1] Berlin-Charlottenburg Bayernallee 28
3) Brandenburg	a)	Suchdienst der Provinzialverwaltung der Mark Brandenburg [2] Potsdam Moltkestraße 6
	b)	Hilfswerk der Evangelischen Kirche in Brandenburg Hauptbüro [1] Berlin-Nikolassee Teutonenstraße 23
4) Mecklenburg	a)	Zentralsuchstelle für Mecklenburg Vorpommern [3] Schwerin Bismarckstraße 10/12
	b)	Hilfswerk der Evangelisch-Lutherischen Landeskirche Mecklenburg Hauptbüro [3] Schwerin Königstraße 19
	c)	Kirchenleitung der Kirchenprovinz Pommern Hauptbüro [3] Greifswald Bahnhofstraße 25/26
5) Land Sachsen	a)	Landessuchstelle für Vermißte beim Polizeipräsidium [10 a] Dresden N Reichspietschufer 2
	b)	Evangelisch-Lutherisches Landeskirchenamt Sachsen Hauptbüro [10 a] Dresden N 6 Rolf-Halm-Straße 1
6) Prov. Sachsen	a)	Präsident der Provinz Sachsen Zentralkartei [19 a] Halle/Saale Trothaer Straße 62
	b)	Vorläufige Kirchenleitung und Evang. Konsistorium der Kirchenprov. Sachsen Hauptbüro [19 b] Magdeburg Am Dom 2
	c)	Evangelischer Landeskirchenrat für Anhalt Hauptbüro [19 b] Dessau-Törten Mösterstraße 53
7) Thüringen	a)	Amtlicher Such- und Meldedienst beim Landesamt für Arbeit und Sozialfürsorge [15] Weimar Karl-Marx-Platz 2
	b)	Hilfswerk der Evangelischen Kirche in Thüringen Hauptbüro [15] Eisenach Pfarrberg 8

Auf Anordnung der sowjetischen Militärverwaltung wurde der Suchdienst für vermißte Deutsche in der sowjetischen Besatzungszone Deutschlands [1] Berlin W 8 Kanonierstr. 35 als Suchzentrale für die gesamte sowjetische Besatzungszone bestimmt. Alle anderen in der sowjetischen Besatzungszone Deutschlands und im sowjetischen Sektor Berlins bestehenden Organisationen werden aufgelöst. Das gesamte Suchmaterial dieser Organisationen wird dem neu geschaffenen Suchdienst übergeben. Für jeden Suchantrag wird eine einmalige Gebühr v. RM 2.— erhoben, welche von dem Suchenden durch Kauf einer Sonderpostkarte bei allen Annahmestellen des Suchdienstes und bei den Postämtern in den Ländern und Provinzen der sowjetischen Besatzungszone zu entrichten ist. Heimkehrer, die jetzt aus der Gefangenschaft zurückkehren, können die Anfrage nach einer Person kostenfrei einreichen.

Der Anschriftensuchdienst der Deutschen Post (10 b) Leipzig C 1 setzt seine Tätigkeit in enger Zusammenarbeit mit dem Suchdienst fort. Anschriften-Meldekarten und Anschriften-Suchkarten sind für 1 Pfg. bei jeder Postdienststelle zu erhalten. Die Gebühr für jede Meldekarte und jede Suchkarte beträgt 50 Pfg. Sie ist auf den Karten in Briefmarken zu verrechnen.

Fortsetzung umseitig

Auslieferung: F. Wittelmeyer (15) Altenburg, Thür., Friedrich-Engels-Str. 19   Preis 0,50 RM.   Genehmigung v. 2. 10. 46.

Mapping motion

Given the whirlwind transformations and great upheavals of the twentieth-century world, the fundamental challenge for cartography was to find new ways to visualise movement in space and change in time, whether the shifting phenomena to be mapped were social structures, humans or features of the natural world. In other words, twentieth-century map makers looked for ways to make maps tell stories.[13]

It was more straightforward, of course, to tell a story using maps. Take, for example, a Chinese communist poster, produced in 1977, a year after the death of China's revolutionary leader Mao Zedong (1893–1976), commemorating his accomplishments as a military leader (ill. 123). The image portrays Mao (on the left) with the future prime minister Chou En-Lai (1898–1976) during the 1940s. They are planning a campaign on a large unfolded sheet-map. Wide, dynamic red-shaded arrows and fainter pencilled circles indicate movements of Red Army forces to surround the enemy. By tracing their journey and plotting its future direction on the map, Mao demonstrates his astute and resolute leadership, his technical expertise (symbolised also by the radio transmitter behind him) and his strategic vision – he uses the map to oversee and command both space and time, as revolution sweeps across and unifies the country, heralding a new era. Movement and change on the map are mirrored by the progression of soldiers through the natural landscape in the background. Maps are instruments of human agency: here, Mao's map does not merely represent but serves to transform China's geography and history. The depiction of the Chairman in this poster was so well received by the Chinese authorities that one of the artists, Liu Wên-his, was commissioned to paint Mao for a new issue of paper currency in 1999.[14]

A poster for Air France, produced in 1950 by Lucien Boucher (1889–1971) as one of his 'Planisphere' series, similarly creates a mythology of movement (ill. 124). It combines a celestial chart, replete with allegorical figures of the constellations, with a globe depicting the company's myriad air routes emanating from Paris. Technological modernity, accelerating time and shrinking distance, is represented by an aeroplane in flight – the sketch is of a Lockheed Constellation, the first airliner with a pressurised cabin. Tradition, universal nature and absolute time and space are signified by the symbolic apparatus of early modern maps – cartouches depicting cherubs blowing out the winds, a compass rose (within which is the Air France emblem, the winged seahorse), the sun's annual path through the twelve signs of the zodiac (two signs are obscured by the globe; Paris is placed with pinpoint accuracy on the sine curve of solar motion). Against the backdrop of France's difficult post-war reconstruction, this poster presented the state airline as symbol and realisation of both progress and tradition, technical innovation and ancient knowledge, speed and timelessness, national pride and universality. It is both a commercial advertisement and patriotic propaganda.

Building on statistical techniques and graphical conventions developed earlier in the field of thematic cartography, twentieth-century mapmakers sought new ways to articulate the time dimension in representations of space.[15] Atlases and other cartographic editions made frequent use of sequences of maps to convey change

121 (previous page)
Three Mile Island Nuclear Station Evacuation Map. Pennsylvania Emergency Management Agency, 1981. Maps CC.5.b.35

122 (left)
Karte der deutschen Flüchtlingssuchstellen, Besatzungszonen und Postleitgebiete. Map of Germany showing zones and postal areas. F. Wittelmeyer, 1945. Maps CC.5.b.52

123 (next page)
Commemorative Chinese poster showing Mao Zedong and Chou En-Lai studying a map in the 1940s. Jên-min ch'u-pan-shê, 1977. Maps 177.f.1.(3)

(see the *National Atlas of Canada*, below), but each map remained a static 'snapshot' of a moment in time. Designers experimented with other methods of incorporating temporality, such as adding transparent paper or plastic overlays; using shading, colouring, timelabels or diagrams; plotting isochrones (lines of equal travel time from a point); employing temporal rather than spatial scales; or including various types of flowlines or arrows.

Designers could do far worse than draw inspiration from physical reality. Flowlines, for example, already exist where they have been inscribed on the earth, sea and sky by objects moving across or through them. These 'maps' in nature conform to a sliding scale of permanence. For example, the vapour trails of aircraft or the waves generated by ships fade away far more quickly than a snail's slime track, a trail left by foxes through grass or a path created by centuries of feet treading the same route. Thus the trade routes and interrelations shown in economic maps from Charles Joseph Minard (1781–1870) onwards, their width calibrated in proportion to the volume of traffic over a specific time period, resemble paths through the landscape that are more heavily and widely worn than others. However, the *duration* and *frequency* of movement or change, with the intervening intervals of stasis – the essential transience of being – are more elusive to capture on a map than the vectors or volume of things in motion.

By the 1960s the traces and imprints of movement in nature were being studied by geographers and artists. Among the latter, Richard Long (1945–) produced work consisting of interventions in the physical landscape, such as his *Line Made by Walking* (1967) (ill. 125). Here, Long photographed a straight line he had worn by purposefully walking back and forth in a field in Wiltshire.[16] The line was at once the material trace of his movement, its inscription in nature and a one-to-one scale map of his mobility; the photograph was its scaled-down representation. In other works he drew his straight, circular or symmetrical walks through the countryside on maps (compare this to Jeremy Wood's work, discussed below). The time taken to 'perform' these walks – their duration – could be gauged by the density of the line or by accompanying lines of text by the artist (ill. 126).

In the Soviet Union, a new state that grounded its claim to legitimacy in notions of social progress and transformation, cartographers expended great effort thinking up fresh ways to communicate the nature and pace of change. The mapping of new spaces and new objects was itself heralded as a sign of progress. Historical mapping contrasted the old with the new. Maps used for planning presaged the future. With a map of the distribution of national minorities in the far north of Russia, published in 1933 but based on data from Arctic expeditions and censuses of the mid-1920s, Soviet ethnographers hoped to create a picture not only of where these tribes lived, but of the scale and extent of their mobility – many were nomadic reindeer herders – and of recent changes in settlement patterns (ill. 127). Population settlements were designated by squares, the size of which was proportional to the number of households. Circles of differing diameters marked the percentage of population groups that were settled in particular areas. Colours differentiated twenty-three nationalities, subdivided into thirty-one smaller ethnicities. In the 1920s the Soviets had collected and mapped this data in

124 (previous page)
Lucien Boucher, Air France poster. Perceval, 1950.
Maps CC.6.a.79

125 (right)
Richard Long, *A Line Made By Walking*, 1967.
Tate P-07149

126
Richard Long, *A Hundred
Mile Walk*, 1971–72.
Tate 1720

Day 1 Winter skyline, a north wind

Day 2 The Earth turns effortlessly under my feet

Day 3 Suck icicles from the grass stems

Day 4 As though I had never been born

Day 5 In and out the sound of rivers over familiar stepping stones

Day 6 Corrina, Corrina

Day 7 Flop down on my back with tiredness
       Stare up at the sky and watch it recede

КАРТА

РАССЕЛЕНИЯ НАРОДНОСТЕЙ
КРАЙНЕГО СЕВЕРА С.С.С.Р.

THE MAP OF DISTRIBUTION OF PEOPLES OF THE FAR NORTH OF U.S.S.R.

СОСТАВЛЕНА ЧЛЕНОМ КОМИТЕТА СЕВЕРА ПРИ ПРЕЗИДИУМЕ ВЦИК П. Е. ТЕРЛЕЦКИМ
ПО ДАННЫМ ПОХОЗЯЙСТВЕННОЙ ПЕРЕПИСИ ПРИПОЛЯРНОГО СЕВЕРА 1926/27 г. И ВСЕСОЮЗНОЙ ПЕРЕПИСИ НАСЕЛЕНИЯ 1926 г.

PREPARED BY P. E. TERLEZKI, MEMBER OF THE COMMITTEE OF THE NORTH FROM DATA OF NEAR-POLAR NORTH HOUSEHOLD CENSUS OF 1926/27 (PART OF THE 1926 CENSUS OF THE UNION).

ПРИ НАНЕСЕНИИ НАСЕЛЕННЫХ ПУНКТОВ БЫЛИ ИСПОЛЬЗОВАНЫ КАРТЫ НАСЕЛЕННЫХ ПУНКТОВ, СПЕЦИАЛЬНО СОСТАВЛЕННЫЕ СТАТИСТИЧЕСКИМИ ОТДЕЛАМИ ИСПОЛКОМОВ — МУРМАНСКОГО ОКРУГА, б. АРХАНГЕЛЬСКОЙ ГУБ., КОМИ АВТОН. ОБЛ., УРАЛЬСКОЙ ОБЛ., б. СИБИРСКОЙ ОБЛ., ЯКУТСКОЙ АССР и ДАЛЬНЕ-ВОСТОЧНОГО КРАЯ и КОМИТЕТАМИ СЕВЕРА б. КИРЕНСКОГО ОКРУГА и ЯКУТСКОЙ АССР, А ТАКЖЕ СХЕМАТИЧЕСКИЕ КАРТЫ ОТДЕЛЬНЫХ ЭКСПЕДИЦИЙ.

В РАБОТЕ УЧАСТВОВАЛИ СТУДЕНТЫ МГУ Б. О. ДОЛГИХ и Я. П. ТЕРЛЕЦКИЙ и ВХУТЕИН'а А. М. КАРЕТНИКОВ.

ИЗДАНИЕ КОМИТЕТА СЕВЕРА ПРИ ПРЕЗИДИУМЕ ВСЕРОССИЙСКОГО ЦЕНТРАЛЬНОГО ИСПОЛНИТЕЛЬНОГО КОМИТЕТА.

PUBLISHED BY THE NORTHERN TRIBES ASSISTANCE COMMITTEE OF THE ALLRUSSIAN CENTRAL EXECUTIVE COMMITTEE.

МАСШТАБ 1:5.000.000.   SCALE 1:5.000.000.

КАРТА
РАССЕЛЕНИЯ НАРОДНОСТЕЙ ЦЕНТРАЛЬНОЙ ЧАСТИ ЯКУТСКОЙ А.С.С.Р.
THE MAP OF DISTRIBUTION OF PEOPLES IN THE CENTRAL PART OF JAKOUT A.S.S.R.
МАСШТАБ 1:1.666.667   SCALE 1:1.666.667

СЕВЕРНОЕ ПОЛЯР

ЗЕМЛЯ ФРАНЦА ИОСИФА

СЕВЕРНАЯ ЗЕМЛЯ

БАРЕНЦЕВО МОРЕ

НОВАЯ ЗЕМЛЯ

КАРСКОЕ МОРЕ

П-ОВ ТАЙМЫР

П-ОВ ЯМАЛ

АРХАНГЕЛЬСК

МУРМАНСКИЙ ОКР.

НЕНЕЦКИЙ НАЦ. ОКР.

ТАЙМЫРСКИЙ (ДОЛГАНО-НЕНЕЦКИЙ) НАЦ. ОКР.

ЯМАЛЬСКИЙ (НЕНЕЦКИЙ) НАЦ. ОКР.

ОСТЯКО-ВОГУЛЬСКИЙ НАЦ. ОКР.

ЭВЕНКИЙСКИЙ НАЦ. ОКР.

ТУРУХАНСКИЙ РАЙОН

СЕВЕРНЫЙ ОКРУГ

ГЛАВЛИТ № 56961. ТИРАЖ 3.000. НАРЯД № 3246.

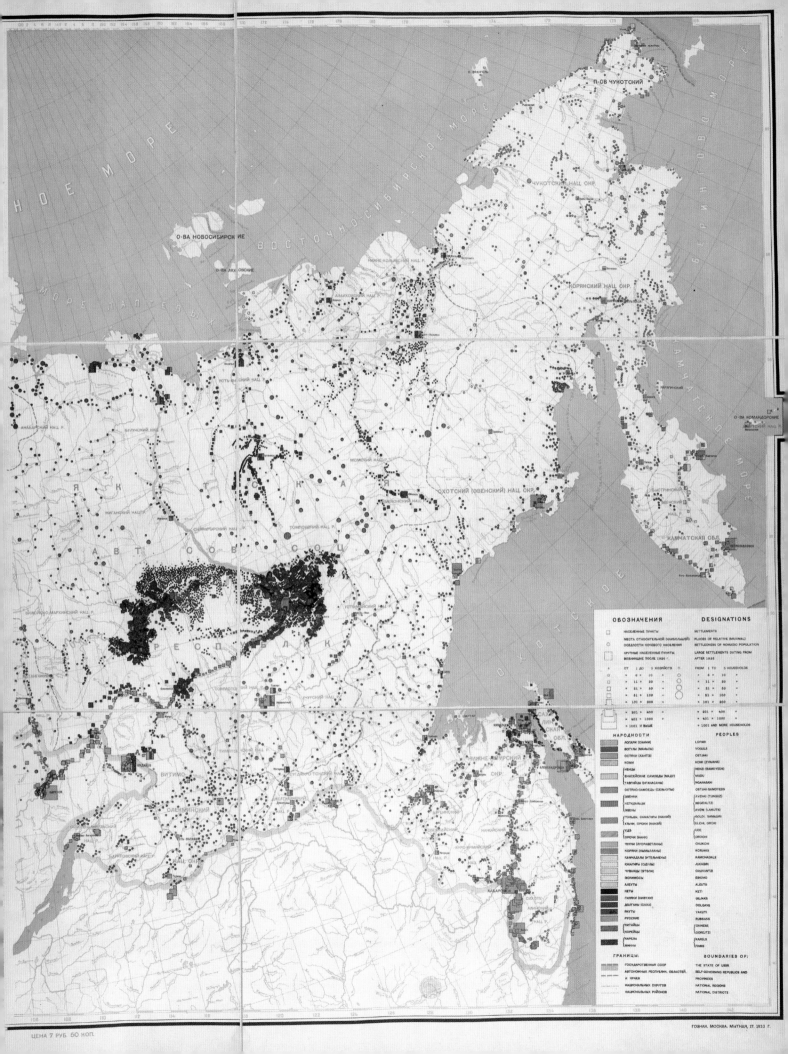

ОБОЗНАЧЕНИЯ   DESIGNATIONS

	НАСЕЛЕННЫЕ ПУНКТЫ	SETTLEMENTS
	МЕСТА ОТНОСИТЕЛЬНОЙ (НАИБОЛЬШЕЙ) ОСЕДЛОСТИ КОЧЕВОГО НАСЕЛЕНИЯ	PLACES OF RELATIVE (MAXIMAL) SETTLEDNESS OF NOMADIC POPULATION
	КРУПНЫЕ НАСЕЛЕННЫЕ ПУНКТЫ, ВОЗНИКШИЕ ПОСЛЕ 1926 Г.	LARGE SETTLEMENTS DATING FROM AFTER 1926

ОТ	1 ДО	5 ХОЗЯЙСТВ	FROM	1 TO	5 HOUSEHOLDS
»	11	20	»	11	20
»	21	50	»	21	50
»	51	100	»	51	100
»	101	200	»	101	200
»	201	400	»	201	400
»	401	1000	»	401	1000
»	1001	И ВЫШЕ	»	1001	AND MORE HOUSEHOLDS

НАРОДНОСТИ   PEOPLES

	ЛОПАРИ (СААМИ)	LOPARI
	ВОГУЛЫ (МАНЬСИ)	VOGULS
	ОСТЯКИ (ХАНТЭ)	OSTJAKI
	КОМИ	KOMI (ZYRIANS)
	НЕНЦЫ	NENZI (SAMOYEDS)
	ЕНИСЕЙСКИЕ САМОЕДЫ (МАДУ)	MADU
	ТАВГИЙЦЫ (НГАНАСАНЫ)	NGANASANI
	ОСТЯКО-САМОЕДЫ (СЕЛЬКУПЫ)	OSTJAK-SAMOYEDS
	ЭВЕНКИ	AVENKI (TUNGUZ)
	НЕГИДАЛЬЦЫ	(NEGIDALTZ)
	ЭВЕНЫ	AVENI (LAMUTS)
	ГОЛЬДЫ, САМАГИРЫ (НАНАЙ)	GOLDI, SAMAGIRI
	УЛЬЧИ, ОРОКИ (НАНАЙ)	ULCHI, OROKI
	УДЭ	UDE
	ОРОЧИ (НАНИ)	OROCHI
	ЧУКЧИ (ЛУОРАВЕТЛАНЫ)	CHUKCHI
	КОРЯКИ (НЫМЫЛАНЫ)	KORIAKS
	КАМЧАДАЛЫ (ИТЕЛЬМЕНЫ)	KAMCHADALS
	ЮКАГИРЫ (ОДУЛЫ)	JUKAGIRI
	ЧУВАНЦЫ (ЭТЭЛЫ)	CHUVANTZI
	ЭСКИМОСЫ	ESKIMO
	АЛЕУТЫ	ALEUTS
	КЕТЫ	KETI
	ГИЛЯКИ (НИВХИ)	GILJAKS
	ДОЛГАНЫ (САХА)	DOLGANS
	ЯКУТЫ	YAKUTI
	РУССКИЕ	RUSSIANS
	КИТАЙЦЫ	CHINESE
	КОРЕЙЦЫ	GOREIZI
	КАРЕЛЫ	KARELS
	ФИННЫ	FINNS

ГРАНИЦЫ: BOUNDARIES OF:

	ГОСУДАРСТВЕННАЯ СССР	THE STATE OF USSR
	АВТОНОМНЫХ РЕСПУБЛИК, ОБЛАСТЕЙ И КРАЕВ	SELF GOVERNING REPUBLICS AND PROVINCES
	НАЦИОНАЛЬНЫХ ОКРУГОВ	NATIONAL REGIONS
	НАЦИОНАЛЬНЫХ РАЙОНОВ	NATIONAL DISTRICTS

ЦЕНА 7 РУБ. 50 КОП.

ГОЗНАК. МОСКВА. МЫТНАЯ, 17. 1933 Г.

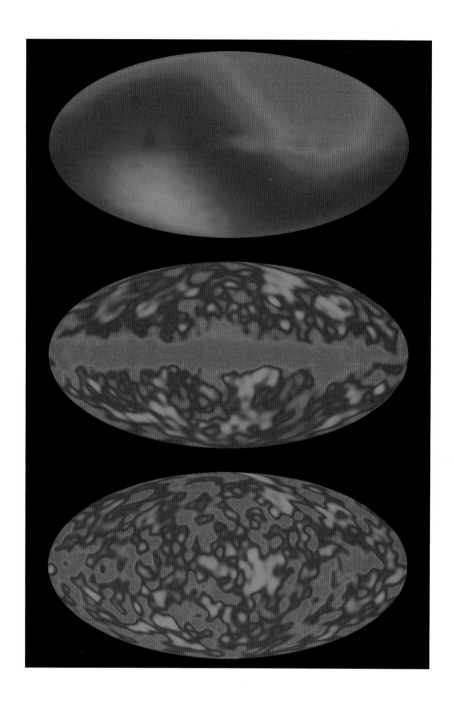

127 (previous page)
Map showing the distribution
of national minorities in the
far north of U.S.S.R. Northern
Tribes Assistance Committee of
the All-Russian Central Executive
Committee, 1933.
Maps 35796.(10)

128 (left)
Cosmic Background Explorer
(COBE). Map of the cosmic
microwave background. Data
collected between 1990 and 1992.
NASA

129 (right)
Heinrich Berann, 'Atlantic Ocean
Floor', *National Geographic
Magazine*, June 1968.
Maps CC.5.b.42

ATLANTIC OCEAN FLOOR

-12000 Depth in feet below sea level   9000 Height above sea level
(14000) Height above the 16,000-foot average depth of the abyssal plains

Produced in the Geographic Art Division
National Geographic Society
MELVIN M. PAYNE, PRESIDENT
for THE NATIONAL GEOGRAPHIC MAGAZINE
MELVILLE BELL GROSVENOR, EDITOR-IN-CHIEF; FREDERICK G. VOSBURGH, EDITOR
WILLIAM N. PALMSTROM, CHIEF, GEOGRAPHIC ART DIVISION

Based on bathymetric studies by Bruce C. Heezen and Marie Tharp of the Lamont Geological Observatory
Painted by Heinrich C. Berann. Compiled by Leo J. Bohanshirski
HORIZONTAL SCALE 1:30,412,800 OR 480 MILES TO THE INCH AT THE EQUATOR
VERTICAL SCALE EXAGGERATED
Mercator Projection
JUNE 1968

order to deliver welfare, healthcare and 'cultural enlightenment' to the nomadic peoples as they moved. During the 1930s the Stalinist regime used this information to promote their settlement; the state's 'sedentarisation' campaigns across the country, especially in Central Asia, frequently entailed coercion and mass violence. In the 1940s German forces in the occupied western Soviet borderlands used similar demographic data collected and mapped by Soviet researchers in the 1920s to identify and isolate targeted populations in those areas. 'In this way', one historian has written, 'Soviet records provided a map for Nazi genocide.'[17] These maps now constitute testimonies to the past, to the impermanence of the human lives and cultures that they mapped and to their own instrumentality in historical change.

As we have suggested, though static maps are able in various ways to represent movement and change, they struggle to convey a sense of duration. They record imprints of the past, sometimes still surviving, often since erased. The maps themselves then become the tracings of lost time. The 1992 Cosmic Background Explorer (COBE), for example, pictured the universe as it 'looked' approximately 400,000 years after the Big Bang, as matter began to clump into what would later become stars, galaxies and clusters of galaxies (ill. 128). It depicted a cosmic geography of the distant past, but was actually a map of tiny fluctuations in thermal radiation as detected by a NASA satellite over two years in the last decade of the twentieth century, as the photons generated nearly 15 billion years ago finally reached our planet, and the satellite's sensors.[18] It is a modern map of traces of deep history.

Similarly, a map of the topography of the ocean seabed by the artist Heinrich Berann (whose Innsbruck Winter Olympics map was discussed above) doesn't illustrate past movement itself but the present-day impression in nature of past movement (ill. 129). The horizontal stress marks running horizontally along the Mid-Atlantic Ridge were fissures caused by seismological movement, the American and African/Eurasian plates moving apart over the course of millions of years. While the similarity between landforms on either side of the Atlantic Ocean had been observed for centuries on maps, the first scientific argument to support the supposition that the earth's crust was mobile was offered by Alfred Wegener (1880–1930) in 1915. But the scientific community largely dismissed his and others' ideas until the 1960s, when separate geophysical and seismological research combined to confirm the theory of continental drift.

Berann's map was one of a number that he designed in the late 1960s based on research that seabed geologists Bruce C. Heezen (1924–77) and Marie Tharp (1920–2006) were doing for the United States Navy at the Lamont–Doherty Geological Observatory (New York).[19] In essence, this was Cold War scholarship, commissioned for strategic purposes – the US Navy required knowledge of deep-sea geology for concealing its nuclear-powered ballistic missile submarines from the Soviet enemy. But Berann's maps popularising the scientists' findings were of huge cultural significance: these rich, striking and beautiful pictures of the uplands and rifts of the ocean floor revealed the hitherto unseen traces left in nature by processes of natural change, prompting many to reconceive the world's physical environment as a unified system, finely balanced, perpetually in flux and motion.

Berann's sea-floor images mapped change from past to present by visualising the extant evidence of movement. Yet maps of the present are, in reality, images of the past. Indeed, we may claim that all static maps that purport to represent the present – other, perhaps, than the most diagrammatic – visualise the past and that, other than as witnesses to history, they are outdated as soon as or even before they are completed. The British Ordnance Survey warns its customers: 'As the landscape is constantly changing, we are unable to survey and map every change as it occurs.' The Ordnance Survey reviews its paper maps at least once every five years; its large-scale topographical datasets are subject to a combination of cyclical and continuous revision.[20]

In 1969 University of Glasgow geographers produced a map of the Breidamerkurjökull glacier in south-eastern Iceland based on aerial surveys conducted by the US Army Map Service in August 1945 (ill. 130). In the interim, as a new 1965 survey had revealed, the finely depicted 'ice edge' of the glacier's snout had in fact receded by a substantial distance and the lagoons at its terminus had expanded. Indeed, further studies showed that the glacier had started to retreat in the 1890s and between 1945 and 1969 receded by between fifty and sixty-two metres per year (although the glacier advanced between 1949 and 1954).[21] The role and importance of the 1969 map based on 1945 data was therefore historical: change and motion could be readily inferred by comparison with a more recent map produced in the mid-1960s.[22] Yet the glacier's continued movement immediately rendered the more recent map out of date. By the end of the century, when Breidamerkurjökull featured in the films *A View to a Kill* (1985) and *Lara Croft: Tomb Raider* (2001; standing in for Siberia), the landscape had shifted and changed even further. Melting ice at the glacier's core now threatened to send it sliding into the fjord. A series of surveys and mappings of this glacier during the twentieth century have contributed considerably to our understanding of climate change.[23]

As we might expect, much swifter natural processes pose even greater challenges to cartographers of the physical landscape. River courses, for example, change rapidly, although all rivers change at different speeds to one another, according to their age and the geological nature of the land they pass through, while human interventions such as reinforcing and embankments can slow this process. Cartographers need to survey and incorporate all such changes into updated maps. Their representations of landscape will always be outpaced by nature. Sometimes, this cartographic time-lag can have political repercussions, such as when changes to river courses complicate their use as markers of state boundaries. A Mexican map produced in 1964 showed changes during the late nineteenth century in a two-and-a-half-mile section of the Rio Grande between El Paso and Ciudad Juárez (ill. 131). Black lines indicated the successive courses of the river and labels marked the dates of their movement. Following the end of the Mexican– American war in 1848, peacemakers had chosen the middle of the river as the boundary between the two countries. But a series of major floods caused the river course to move southwards by over two kilometres by 1910, giving rise to a long-running diplomatic row over ownership of the 600-acre Chamizal tract of land that had shifted from the Mexican side of the river to the US side.[24] International tensions were further aggravated when joint Mexican–US measures to prevent further flooding inadvertently created an island

130 (overleaf)
Breidamerkurjökull South East Iceland, August 1945.
University of Glasgow, 1969.
Maps Y.39

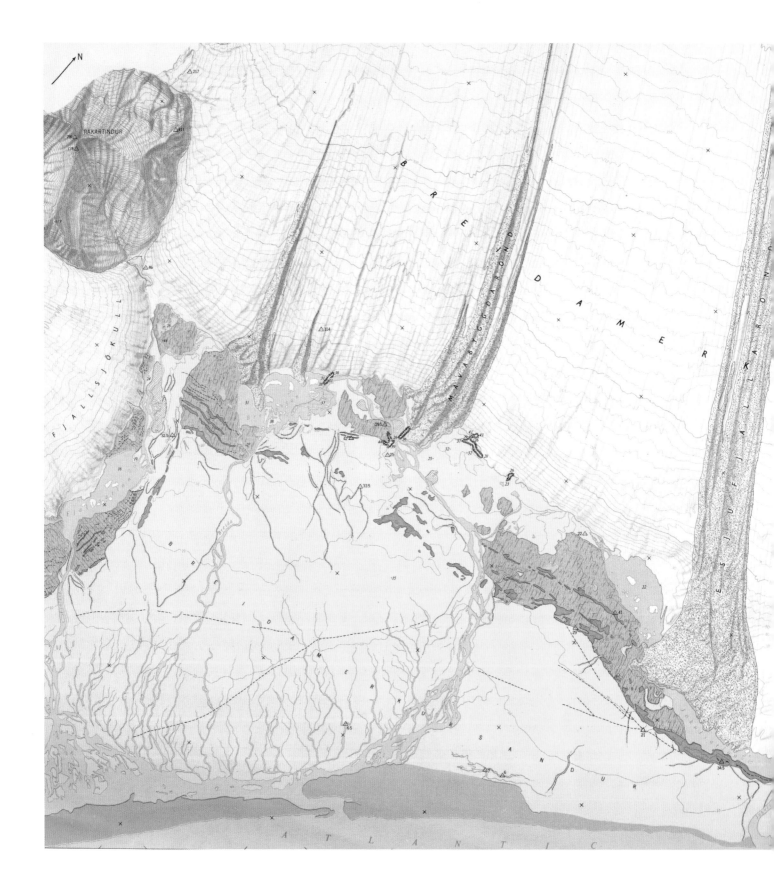

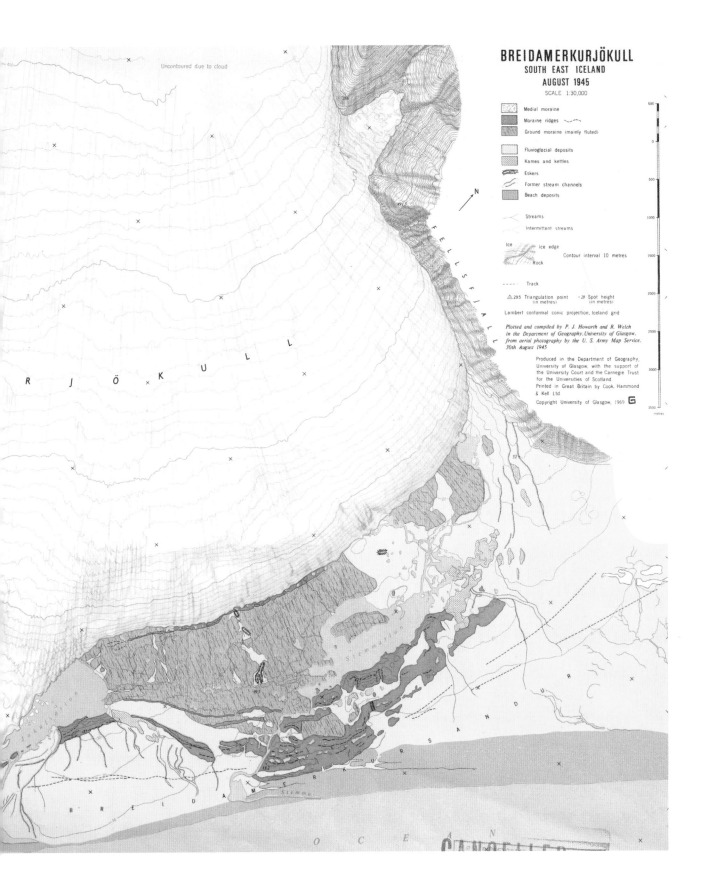

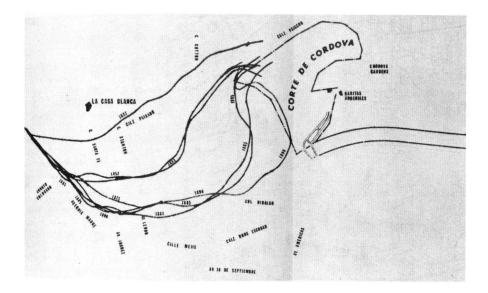

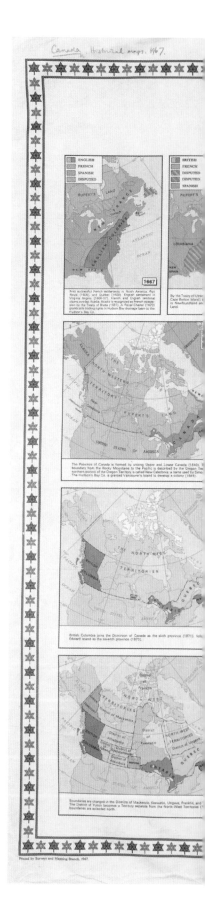

in the river (marked 'Corte de Cordova' on the map). The island was legally Mexican but now situated within US territory, and it became a no-man's land notorious for uncontrolled border crossings, drug trafficking, illicit alcohol trading and violence.[25] This map was one of a number celebrating the resolution of the dispute in 1964, when the USA returned the Chamizal tract to its southern neighbour, and its American population were 'repatriated' to US territory.

In the Chamizal dispute, historical maps were used to justify and describe Mexico's legal claim to the land as well as in reaching a settlement. Historical maps can also play a significant role in creating the mythologies of landscape, home and community in which nations root their sense of identity. For some, the national story is one of territorial expansion, the shifting frontier and 'manifest destiny'. For others, it is a tragedy of spatial disintegration and loss (see the discussion of 'Hungaria' below). A map compiled in 1967 for inclusion in the *National Atlas of Canada* contains twenty-three smaller maps of North America showing successive borders in chronological sequence from 1667 to 1949 (ill. 132). This 'multiple map strategy' enabled the viewer to compare each map with the next and to perceive the changes that had taken place over time.[26] The map was designed to give the impression that Canada's acquisition of territory had been a just, inevitable and smooth process. Like many such atlases, the *National Atlas of Canada* was an ideological as much as as intellectual exercise, aiming to promote patriotism by disseminating a particular picture of national geography and history. In such a context, this map functioned to legitimate Canadian statehood by making a visual analogy between its territorial evolution and the apparent inevitability of geological movement.

131 (above)
Map showing the changing course of the Rio Grande, from *El Chamizal, monumento a la justicia internacional*. Secretaria de Hacienda, 1964.
X.700/1229

132 (right)
Territorial Evolution of Canada. Department of Energy, Mines and Resources, 1967.
Maps 70619.(140)

ᴇRRITORIAL EVOLUTION OF CANADA

Scale 1:50,000,000 or 797 Miles to One Inch.
THESE MAPS ARE TO BE PUBLISHED IN THE ATLAS OF CANADA

Produced by the Geographical Branch

Copies may be obtained from the Map Distribution Office, Department of Energy, Mines and Resources, Ottawa, Canada.

Mobile mapping

In the last section, we looked at some ways in which twentieth-century cartographers sought to represent movement and change. We suggested that the static map, as a unified view created at a particular moment in time and from a fixed perspective, is unable to convey adequately any sense of the duration or tempo of transformation or mobility. In other words, maps are not conventionally good at telling stories. But maps — all maps — represent time as much as space: they show not the present but traces of the past or visions of the future. In this sense, almost all traditional maps are always either obsolete, other than as witnesses or instruments of history, or imaginary.

During the twentieth century, however, new technologies enabled cartographers to make maps that were dynamic (i.e. moving and changing) and that integrated mobile perspectives (i.e. represented shifting viewpoints, often in relation to the movement of the user). This held out the possibility of maps that represented motion in real time and could continuously update themselves.

A postcard map, *Hungaria 896–1918* (ill. 133), is a rudimentary, though ingenious, early example of a dynamic map. Produced by the Hungarian Women's National Association in 1920, it depicted the dismemberment of defeated Hungary in the post-First World War territorial settlement. Under the Treaty of Trianon of June 1920, Hungary lost over two-thirds of its former territory, as well as over half of its population. One-third of Hungary's Magyar speakers found themselves in foreign countries — Czechoslovakia, Romania, Yugoslavia or Austria. The settlement created tensions that destabilised Europe in the interwar years and that continue to cause ructions in the present day. By turning a cogwheel on the side of the postcard, the user 'exploded' into separate parts a map showing what nationalists claimed as the ancient, inalienable Hungarian lands, revealing the form and extent of its territorial losses. By turning the wheel in the opposite direction, the user could bring about the reunification of the nation.[27] The postcard was designed not merely to communicate spatial information 'before' and 'after' change (as a map series might do — see *National Atlas of Canada*, above). By using primitive animation, and by engaging the viewer not only visually but also physically in its operation, the map first and foremost was a propaganda tool that aimed to generate an emotional response. The inclusion of territorial and demographic data drives its message home.

Innovators dreamed up other methods to make maps move or to simulate movement. As car ownership expanded in the 1920s in Europe and America, manufacturers designed a number of mechanical gadgets to aid navigation that used rollers or levers to turn or flip maps more smoothly than manual page-turning. One such device, the German *Karten-Wunder* of the late 1930s, comprised a small Bakelite cassette, the size of a paperback book, containing thirteen multicoloured map sheets, each made up of nine sliding panels, and together covering the whole territory of the country and featuring Hitler's newly built *Autobahn* network (ill. 134). The user switched from one map to the next by pressing tabs on the side of the case. Ultimately, of course, these devices remained constrained by the inert nature of the conventional map.

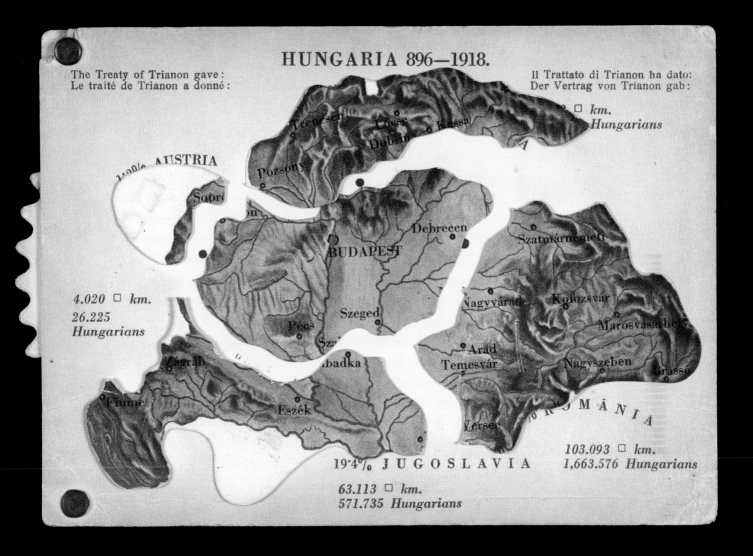

HUNGARIA 896—1918.

The Treaty of Trianon gave:
Le traité de Trianon a donné:

Il Trattato di Trianon ha dato:
Der Vertrag von Trianon gab:

□ km.
Hungarians

1·90% AUSTRIA

4.020 □ km.
26.225
Hungarians

103.093 □ km.
1,663.576 Hungarians

19·4% JUGOSLAVIA

63.113 □ km.
571.735 Hungarians

133
Hungaria 896–1918. Postcard by
Emich. Hungarian Women's National
Association, 1920.
Maps C.12.c.1.(807)

134 (left)
Karten-Wunder. Autobahn road map device, produced in the late 1930s. Maps 188.v.39

135 (above)
Screenshot from *Casablanca*, directed by Michael Curtiz, 1942

As the century progressed, more sophisticated technologies enabled cartographers to capture and convey the time dimension in ways that were both subtle and powerful. Cinema, in particular, gave people a new way of looking at reality that not only could record change in time and movement in space, but by means of dynamic camerawork and editing could communicate the perception or experience of flow and flux. Scholars have argued that many cinematic techniques not only anticipated but shaped future innovations in dynamic cartography.[28]

Moving pictures had featured animated maps since the start of the twentieth century, as a way of identifying the space of the action, interrelating different places and illustrating movement between them. In feature films, maps also functioned to create an illusion of documentary authenticity. One of the most famous examples of the use of maps in film is the opening sequence of the 1942 American movie *Casablanca*, directed by Michael Curtiz (ill. 135). As the credits roll, the viewer sees a static political map of colonial Africa in the background – the world as it was. This is followed by the image of a slowly rotating relief globe, placing the story to be told in the context of the world war, and linking the Pacific theatre (the concern of most US audiences at the time) with Europe and North Africa (the scene of the story). The camera zooms in on Western

231

Europe, then on France. The close-up of the globe dissolves into a close-up of a map centred on Paris. The camera then zooms out to show a map of France with a refugee route to Marseille indicated by an animated line extending southwards, superimposed on newsreel footage of people in flight. As the route crosses the Mediterranean, the solid line becomes a dotted line and the newsreel photos show a large ship. The camera follows the line as it extends southwards, panning down and closing in on North Africa, where it becomes solid again as it winds to the south-west, again with film footage of refugees in the background. Finally the line terminates in Casablanca, and the map fades out into a postcard-like view of a minaret and palm tree overlooking the city and sea. The camera then pans down to bustling street level, where the film's action will take place. Throughout this sequence, dramatic music plays and a male voice-over describes the routes taken by refugees fleeing France, and the dangers, difficulties and decisions they faced. The blending of sound, documentary footage and maps works to convey both information and atmosphere to the audience, and to integrate history and fiction in a single audiovisual narrative of events, movement, space and place.[29]

In several ways, this sequence established in Western popular consciousness and culture ways of viewing the world through maps that have become familiar in recent years to users of Google Earth: the zoom from the global to the local ('scale animation'); the mobile viewpoint ('fly-through'); the 'jump' between scales; changes in perspective ('tilt effect'); and the switch from map to photograph ('street view'). Indeed, a direct lineage can be traced between the cartographic artistry of *Casablanca*, through the 1977 animated film by Charles and Ray Eames called *Powers of 10* (that consisted of a smooth zoom-out from a close-up of a picnic in a Chicago park to the universe, during which the viewpoint recedes by a power of ten every ten seconds, and then a zoom-in to the sub-atomic scale), to the software that was developed into Google Earth.[30]

Inspired by cinema's use of maps, the University of California scholar Norman Thrower wrote an influential article in 1959 urging fellow cartographers to adopt animation as a way of mapping change and motion in society and nature.[31] Within a decade, the revolution in computing had opened up new possibilities for rendering ever more sophisticated and seamless animations. As digital technologies developed further, designers were increasingly able to produce maps that provided their users with integrated and automatic mobile perspectives.

In 1981, for example, the car manufacturer Honda introduced the Electro Gyrocator as an option for its cars. This navigation system required drivers to insert street maps printed onto sheets of film into a reader that, using a gyroscope, accelerometer and microcomputer, displayed the vehicle's movement on the map in real time. Other early electronic navigation devices for cars combined inertial systems with on-board geospatial databases that could redraw the map as the vehicle moved.[32] Around the same time, companies such as Racal-Decca Navigation produced systems for aeronautical navigation that used strip maps printed on paper or film that moved on rollers controlled either electronically or by computer. Some of these devices had a moving pointer that indicated the aircraft's ground location on the map. On others, the map was optically projected onto a cockpit display panel.[33]

136
Avidyne moving map navigation
systems from the 1990s

Case Study: 3D Displays of Internet Traffic, K.C. Cox and S.G. Eick, pp. 129-131.

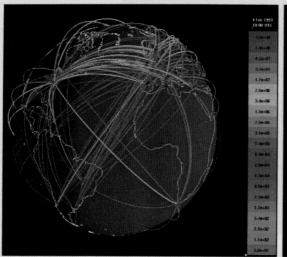

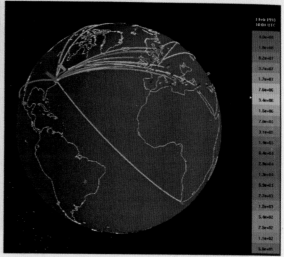

Figure 3: One frame from a animation showing worldwide internet traffic over a two hour period.

Figure 5: Thresholding the link statistics with a slider (not shown) highlights the most important links and helps minimize display clutter.

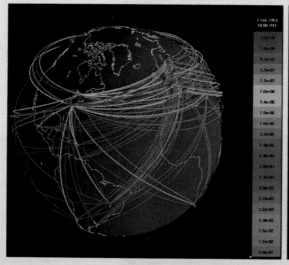

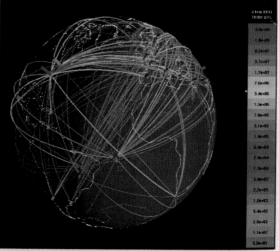

Figure 4: Rectangular latitude-longitude paths.

Figure 6: Making the globe translucent reveals obscured arcs.

152

To establish location, some of these new aeronautical and maritime navigation systems tracked radio signals broadcast from chains of beacons. Others used radar, which since the Second World War had offered pilots and ships' captains, air-traffic controllers, operators of early-warning systems, meteorologists and many others a mapping of the position and motion of objects relative to a central point (the radar antenna). The Avidyne Flight Situation Display introduced in the 1990s, for example, combined full-colour radar, its adverts claimed, with 'the most sophisticated moving maps available, digitized charting displays, and enhanced Stormscope® displays [providing weather information] with colour contouring' (ill. 136). The company's tagline was: 'Avidyne. Because we not only see the future of situational awareness. We are the future.'[34]

The twin development of aviation and photographic technologies in the early twentieth century gave artists, cinema-goers, urban planners, cartographers and others a new way of looking at the world and a new perspective on social reality.[35] Many scholars and cultural critics have argued that the aerial view, with its abstracted, dehumanised, 'gridded' and ordered vision of society and nature, is a fundamental constituent of modern culture and politics.[36] Initially, during the First World War, aerial survey was used for military reconnaissance and mapping. In the following decades, it was adapted for civilian cartography. Aeroplanes equipped with photographic apparatus enabled faster surveying of new territories and a more efficient updating of existing maps.[37]

Pilots conducting aerial surveys had to fly with absolutely precise speed and bearing. On the ground, technicians then took the photographic plates or film – each frame a static snapshot taken by a fast-moving craft – and, using techniques of photogrammetry, telemetry and stitching, melded them together into smooth and contiguous maps. Since the 1960s, near-orbit satellites have carried out mapping of even wider areas at ever higher resolutions, their sensors using the full range of the spectrum to reveal patterns and processes invisible to the eye.[38] Remote sensing thus generates sequential, shifting images that are unified both temporally and spatially on the resulting map. In this sense, the single fixed perspective of the modern topographical map is as much an illusion as its temporal 'presentness'.

The creation of the World Wide Web in the last decades of the twentieth century opened up an entirely new landscape that was itself abstract and dynamic. This network required moving maps to describe it as well as offering a new means to deliver mobile mapping to users. The Internet is made up of data flows between servers and between users and servers: users send requests for information (or 'hits') to a server, which responds in kind. The complexity of operations and the necessity of monitoring the flow of data traffic through cables and via satellites (much like road traffic) necessitated the creation of moving data maps. One such map, produced by Stephen Eick and Kenneth Cox at Bell Laboratories in 1995, visualised the volume and location of Internet traffic over a set time period (ill 137).[39] Another was the 1998 WebPath produced by Emmanuel Frécon and Gareth Smith, a 'surf map' which dynamically and in three dimensions visualised a user's navigation through the Internet.[40] These data flows mapped and constituted a new space, which, albeit virtual, was in many ways as real as physical space.

137

3-D Displays of Internet Traffic,
K.C. Cox and S.G. Eick, CA.
Institute of Electrical and Electronic
Engineers Computer Society Press,
1995
4496.403000

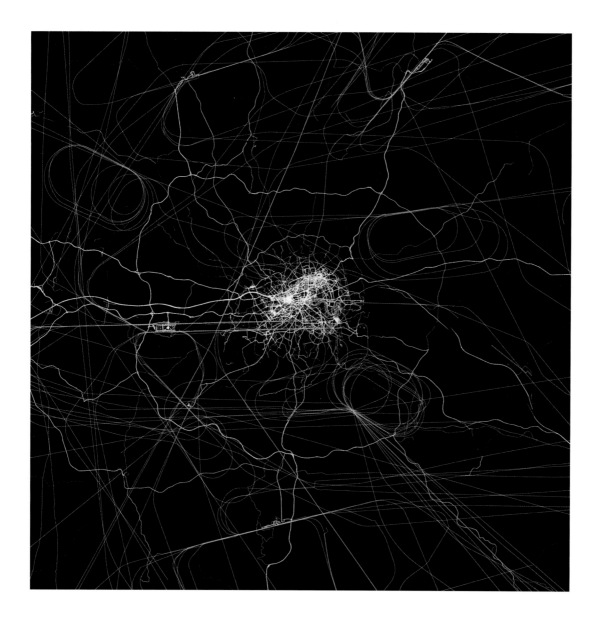

138
Jeremy Wood, *My Ghost*, 2016.
Maps CC.6.a. 83

The technology which finally placed the car, ship or aircraft and, from the 1990s, the individual automatically at the centre of the moving map, kick-starting the subject-centred mapping revolution of the twenty-first century, was the Global Positioning System (GPS). The US Department of Defense launched development of Navstar GPS in 1973 (the first satellite was launched in 1978), in parallel with the rival Soviet GLONASS. These (and later European and Chinese networks) used continuous, precisely timed signals broadcast by orbiting satellites to fix the locations and thus map the movement of receivers.[41] One historian has described global navigation satellite systems (GNSS) as the culmination of technological developments in the twentieth century that 'tied clocks and maps ever closer together'.[42] GPS has transformed cartography, military strategy, navigation and time synchronisation (it is used, for example, in the coordination of traffic-light systems) and has made possible the real-time tracking of multiple mobilities, including of wildlife, parcels and people.

Since its diffusion among civilian users – in 2010 about 1 billion GPS receivers were in use worldwide, a number which has since grown vastly with the integration of GPS into smartphones – real-time dynamic mapping has begun to reshape our reality in profound ways. As one scholar has written, 'Both the spaces of day-to-day experience and the spaces constructed by representational maps were superseded by a space that was more immediately calculable, less historical, and almost perfectly uniform.'[43] In this conception, twenty-first-century mapping has further dematerialised the modern map, which Harvey (cited at the start of this chapter) saw as representing 'abstract and strictly functional systems for the factual ordering of phenomena in space'.[44] At the same time, maps and mapping practices, such as walking through the landscape or navigating city streets, are reunified through dynamic digital mapping. But new technologies, in bringing the map to life, are encouraging, if not forcing, us to live 'inside the map'. A new consciousness of our relationship with the world emerges: we are moving = we are mapping = we are being mapped.

Seeking to understand the impact of new technologies on our conceptions of the environment, of ourselves and of our place in it, artists such as Jeremy Wood (1976–) have used GPS to record the movement of golf-course lawnmowers and to write messages on the landscape (referencing Richard Long's *Line Made by Walking*, discussed above). One of Wood's best-known creations is a series of maps visualising his movement in and around London over periods of years, as tracked, recorded and drawn by GPS and mapping software (ill. 138). Thes artworks, all entitled *My Ghost*, are representations of his 'GPS footprint' – they render his transient motion as permanent marks. The frequency of his walks along certain routes is shown by the brightness and density of the lines, as the digital tracks of his wanderings, undertaken at different times, are superimposed on each other in a single static image. With this artwork, Wood returns us to the notion of the map as a picture of traces of the past. With this, he has also taken ownership of his own space and his own history.

Conclusion

For most of the twentieth century, maps were static pictures of bounded space, unable – despite the design innovations surveyed earlier in this chapter – fully to integrate temporality or to represent the movement or transformation of things in time. At the century's end, the technological revolution, discussed in the preceding section, enabled maps to become dynamic, mobile, open and interactive, representing – often synchronously – the very processes and experiences that made them. As such, they have begun to exert an ever more powerful influence over our perceptions and practices. Maps now determine what is possible – insofar as they reveal opportunities – and what is permissible – insofar as they establish what Deleuze and Guattari (cited earlier) termed 'fixed paths in well-defined directions'.[45] Ubiquitous and constant mapping, on our satnavs, mobile phones and smartwatches, has become part of the modern way of life, for better and for worse.

Using mobile mapping devices, we track and record our own trajectories in real time and create spatialised narratives of ourselves that we share with others via social media. We are sure to 'check in' when we visit somewhere exciting or exotic. We add pictures as soon as we take them. We leave our digital tracks everywhere as we move through both the online and the physical worlds. In our travels, and at home, we are always at the centre of our own map, even as the map – like the route maps discussed at the start of this chapter – constrains and regulates our mobility and behaviour. In this way, our 'situational awareness' becomes about not merely *where* we are but also *who* we are. Our new virtual – but no less real – personas have merged with or even supplanted the identities we create for ourselves through everyday physical interactions.

We are all cartographers now. Or, at least, so it seems. In fact, most of the interactive dynamic maps that we use to track, record and represent our movements have been created to make visible to us what others – corporations or governments – want or permit us to see: a selection of features, amenities, services and products, and the routes that lead us to them. As such, these maps are no less functional and instrumental than conventional cartography, although they are far more persuasive and powerful in the ways they condition our everyday lives, by substituting representation for reality while creating the twin illusions of agency and authenticity. Critics have warned that GPS combined with ubiquitous interactive mapping based on (and feeding) ever vaster geographic information systems (GIS, i.e. geospatial databases such as Google Earth and Google Maps) establish the conditions for a new 'geoslavery'.[46]

While we are mapping, and being mapped, we are at the same time losing ourselves. Not in the traditional sense, when we wandered 'off the map' and found out something new, about a place or about ourselves.[47] Nor even in the sometimes farcical, sometimes tragic sense, when satnav misdirection leaves lorries swamped in streams, cars overhanging cliff edges or cruise ships run aground.[48] We are losing the knowledge, ability and will to develop and exercise a sense of direction; to orient ourselves autonomously in the world; to engage, on our own terms, with unmapped, marginal or 'accidental' geographies and histories; and to grasp intuitively the intimate and necessarily two-way interrelationship between ourselves and the natural environment. It is in these ways that we risk, collectively and individually, forgoing an essential part of our nature as humans.

As early as 1995, the scholar and critic John Pickles expressed disquiet over new mapping technologies in a collection of essays (prepared in collaboration with the distinguished map historian J. B. Harley) entitled *Ground Truth: The Social Implications of Geographic Information Systems*. While acknowledging the many advantages of GPS and GIS, Pickles voiced concern about 'unmediated technical practices' and emphasised a need 'to be clear about the different ways in which technologies affect specific groups and regions, reconfigure social relations, and increase the potential for the exploitation of some to the benefit of others'.[49] Yet, since this warning, efforts to establish safeguards protecting citizens' privacy, and to promote public awareness of the potential dangers of new technologies and techniques of geolocation and mapping, have been continuously outpaced and often superseded by innovation.[50]

Maps curate our experience of the world as well as guiding our movement through it. We need to understand better how maps have functioned in the past and how they operate now, in this era of interactive, mobile mapping. And we need to reflect on whether we wish our lives to be curated by cartographic technologies harnessed to the interests of commerce or social control, or by community and citizen mapmakers (that is, by each and every one of us), who are able – with only a little investment of time and effort – to take advantage of powerful non-proprietary platforms, such as OpenStreetMap, to reclaim our public spaces and preserve our private selves.[51]

NOTES

Introduction: Drawing the Line

1. This introduction has benefitted enormously from the thoughts and advice of Peter Barber, Nick Baron and Michael Heffernan.

2. The edict that 'a map is not the territory it represents', made by Alfred Korzybski in 1933, was an early statement in a discussion that has been revisited up to the present day. *Science and Sanity: An Introduction to Non-Aristotelian Systems and General Semantics* (Lancaster, PA: International Non-Aristotelian Library Publishing Co., 1933), p. 58.

3. Stuart Elden, *The Birth of Territory* (Chicago: University of Chicago Press, 2013), p. 322.

4. John Pickles, *A History of Spaces: Cartographic Reason, Mapping, and the Geo-coded World* (London: Routledge, 2004), p. 40.

Chapter 2. Maps and Peace

1. Tim Bryars and Tom Harper, *A History of the Twentieth Century in Maps* (Chicago: University of Chicago Press, 2014); Mark Monmonier, *The History of Cartography, vol. 6: Cartography in the Twentieth Century* (Chicago: University of Chicago Press, 2015).

2. Timothy Barney, *Mapping the Cold War: Cartography and the Framing of America's International Power* (Chapel Hill, NC: University of North Carolina Press, 2015); Guntram Hendrik Herb, *Under the Map of Germany: Nationalism and Propaganda 1918–1945* (London: Routledge, 1996); Susan Schulten, *The Geographical Imagination in America 1880–1945* (Chicago: University of Chicago Press, 2001).

3. James R. Akerman (ed.), *The Imperial Map: Cartography and the Mastery of Empire* (Chicago: University of Chicago Press, 2009); Catherine Dunlop, *Cartophilia: Maps and the Search for Identity in the French–German Borderland* (Chicago: University of Chicago Press, 2015); Matthew Edney, 'The Irony of Imperial Mapping', in Akerman, *The Imperial Map*, 2009, pp 11–45, 303–16.

4. See, more generally, Jay Winter, *Dreams of Peace and Freedom: Utopian Moments in the Twentieth Century* (New Haven, CT: Yale University Press, 2006).

5. Denis Cosgrove, *Apollo's Eye: A Cartographic Genealogy of the Earth in the Western Imagination* (Baltimore, MD: Johns Hopkins University Press, 2001).

6. G. S. Dunbar, 'Elisée Reclus and the Great Globe', *Scottish Geographical Magazine*, vol. 90, no. 1 (1974), pp 57–66.

7. H. G. Wells, *The War that Will End War* (London: Frank & Cecil Palmer, 1914).

8. Denis Cosgrove, *Apollo's Eye*, pp 257–62; Robert Poole, *Earthrise: How Man First Saw the Earth* (New Haven, CT: Yale University Press, 2010).

9. Alastair Pearson, D. R. Fraser Taylor, Karen D. Kline and Michael Heffernan, 'Cartographic Ideals and Geopolitical Realities: International Maps of the World from the 1890s to the Present', *Canadian Geographer*, vol. 50, no. 2 (2006), pp 149–76.

10. Michael Heffernan, 'Geography, Cartography and Military Intelligence: The Royal Geographical Society and the First World War', *Transactions of the Institute of British Geographers*, vol. 21, no. 3 (1996), pp 504–33.

11. Alastair Pearson and Michael Heffernan, 'The American Geographical Society's Map of Hispanic America: Million Scale Mapping between the Wars', *Imago Mundi*, vol. 6, no. 2 (2009), pp 1–29.

12. Alastair Pearson and Michael Heffernan, 'Globalizing Cartography: The International Map of the

World, the International Geographical Union, and the United Nations', *Imago Mundi*, vol. 67, no. 1 (2015), pp 58–80.

13. Arthur Robinson, 'The Future of the International Map', *Cartographic Journal*, vol. 1 (1965), pp 1–4.

14. Jeremy Crampton, 'Cartography's Defining Moment: The Peters Projection Controversy, 1974–1990', *Cartographica*, vol. 31, no. 4 (1994), pp 16–32; Mark Monmonier, *Rhumb Lines and Map Wars: A Social History of the Mercator Projection* (Chicago: University of Chicago Press, 2004).

15. Duncan Campbell, *War Plan UK: The Secret Truth about Britain's 'Civil Defence'* (London: Paladin, 1982).

16. Stan Openshaw, Philip Steadman and Owen Greene, *Doomsday: Britain After Nuclear Attack* (Oxford: Blackwell, 1983).

17. William Bunge, *Nuclear War Atlas* (Oxford: Blackwell, 1988).

18. Timothy Barney, *Mapping the Cold War*, pp 170–214.

19. www.visionofhumanity.org [accessed 26 July 2016].

Chapter 3. Everyday Maps

1. Joseph Conrad, *Heart of Darkness*, in *Youth: A Narrative, and Two Other Stories* (Edinburgh: William Blackwood & Sons, 1902).

2. https://www.ordnancesurvey.co.uk/about/news/2015/os-acquire-shareholding-in-dennis-maps.html [accessed 26 July 2016].

3. Tim Bryars and Tom Harper, *A History of the 20th Century in 100 Maps* (London: British Library, 2014), p. 177.

4. *The Publisher's Circular and Booksellers' Record,* vol. 98 (London: Publisher's Circular, 1913), p. 96.

5. To quantify the purchasing power of specific amounts, see www.measuringworth.com.

6. D. P. Bickmore and M. A. Shaw (eds), *The Atlas of Britain and Northern Ireland* (Oxford: Oxford University Press, 1963).

7. David Leboff and Tim Demuth, *No Need to Ask! Early Maps of London's Underground Railways* (Harrow Weald, Middlx: Capital Transport, 1999), p. 33.

8. Hazel Sheeky Bird, *Class, Leisure and National Identity in British Children's Literature, 1918–1950* (Basingstoke: Palgrave Macmillan, 2014), p. 89.

9. Bryars and Harper, *History*, p. 70.

10. Rudyard Kipling, *Just So Stories for Little Children* (London: Macmillan, 1902), pp 104–5.

11. http://bearalley.blogspot.co.uk/2010/02/ronald-lampitt.html [accessed 26 July 2016].

12. *Journal of Education*, vol. 80 (1948), p. 714.

13. Rex Walford, *Geography in British Schools 1850–2000: Making a World of Difference* (London: Woburn Press, 2001), p. 95.

14. Laurie Lee, *Cider with Rosie* (London: Hogarth Press, 1959), p. 63.

15. Pamela Horn, *The Victorian and Edwardian Schoolchild* (Gloucester: Alan Sutton, 1989), p. 167.

16. Walford, *Geography in British Schools*, p. 72.

17. Robert Stephenson Smyth Baden-Powell, *Scouting for Boys: A Handbook for Instruction in Good Citizenship* (London: C. Arthur Pearson, 1908), p. 37.

18. Baden-Powell, *Scouting for Boys*, p. 194.

19. Michael J. Childs, *Labour's Apprentices: Working Class Lads in Late Victorian and Edwardian England* (Montreal: McGill-Queen's University Press, 1992), p. 141.

20. Baden-Powell, *Scouting for Boys*, p. 378.

21. Bryars and Harper, *History*, pp 20–1.

22. Cecil H. Crofts, *Britain On and Beyond the Sea: Being a Handbook to the Navy League Map of the World*, 5th ed. (Edinburgh and London: W. & A. K. Johnston, 1909), p. v.

23. Crofts, *Britain On and Beyond the Sea*, p. vii.

24. Childs, *Labour's Apprentices*, p. 155.

25. Bryars and Harper, *History*, p. 27.

26. *Manual of Map Reading and Field Sketching* (London: HMSO, 1914), p. 5.

27. Peter Chassaud, *Mapping the First World War* (Glasgow: Collins, 2013), p. 9.

28. T. H. Hawkins and L. J. F. Brimble, *Adult Education: The Record of the British Army* (London: Macmillan, 1947), pp 42–3.

29. E. W. Hornung: *Notes of a Camp Follower on the Western Front* (London: Constable, 1919), p. 96.

30. Hornung, *Notes of a Camp Follower,* p. 108.

31. George MacDonald Fraser, *The Sheikh and the Dustbin and Other McAuslan Stories* (London: Collins and Harvill, 1988).

32. Spike Milligan, *Adolf Hitler: My Part in his Downfall* (London: Michael Joseph, 1971).

33. Michael Burleigh, *Moral Combat: A History of World War II* (London: Harper Press, 2010), p. 360.

34. Hawkins and Brimble, *Adult Education,* pp 120–1.

35. Hawkins and Brimble, *Adult Education,* p. 245.

36. Hawkins and Brimble, *Adult Education,* p. 361.

37. Jeremy A. Crang, *The British Army and the People's War 1939–45* (Manchester: Manchester University Press, 2000), p. 119.

38. Museum of London item: MoL_73.240/19a and Imperial War Museum item: EPH 10930.

39. Angela Brazil, *The Luckiest Girl in the School* (London: Blackie & Son, 1916), p. 154.

40. *Air Raid Map of the Metropolitan Area and Central London; Air Raids and Naval Bombardments,* in J. A. Hammerton (ed.), *Harmsworth's New Atlas of the World – Supplement* (London: Amalgamated Press, *c.* 1919–20).

41. *Map of the Bombing of Canterbury, reprinted in aid of RAF pilots and crews fund,* printed at the Office of 'The Kentish Gazette & Canterbury Press', [s.l. c. 1944].

42. Michael Heffernan, *The Cartography of the Fourth Estate: Mapping the New Imperialism in British and French Newspapers c. 1875–1925,* in James Akerman (ed.), *The Imperial Map: Cartography and the Mastery of Empire* (Chicago: University of Chicago Press 2009), p. 269.

43. Heffernan, *Cartography,* p. 270.

44. Bryars and Harper, *History,* p. 149.

45. *Illustrated London News Election Map,* in *Illustrated London News Supplement,* 13 January 1906.

46. Alexander Radó, *The Atlas of To-day and To-morrow* (London: Victor Gollancz, 1938), p. ix.

47. Michael Heffernan and Robert Győrii, *Sándor Radó (1899–1981), Geographers: Bibliographical Studies, vol. 33* (2014), p. 179.

48. Heffernan and Győrii, *Sándor Radó,* p. 179.

49. John Feather, *A History of British Publishing,* 2nd edn. (London: Routledge, 2005), p. 157.

50. Feather, *History of British Publishing,* pp 167–8.

51. James Francis Horrabin, *The Atlas of Current Affairs,* 6th impression (London: Victor Gollancz, 1936), p. 5.

52. Horrabin, *Atlas,* pp 80-1.

53. Bryars and Harper, *History,* pp 78–9.

54. J. J. Cherns, *Official Publishing: An Overview* (Oxford: Pergamon Press, 2013), pp 254–5.

55. 'It is quite likely that this information hasn't been retained as only 5% of government documents are selected for permanent preservation at The National Archives' (correspondence, January 2016).

56. Patrick Abercrombie, *Greater London Plan 1944* (London: HMSO, 1945); British Railways Board, *The Reshaping of British Railways* (London: HMSO, 1963); Colin Buchanan, *Traffic in Towns: A Study of the Long Term Problems of Traffic in Urban Areas* (London: HMSO, 1963).

57. E. J. Carter and Ernő Goldfinger, *The County of London Plan Explained* (London: Penguin, 1945).

58. Bryars and Harper, *History,* p. 111.

59. Bryars and Harper, *History,* p. 212.

60. Otto Neurath (Matthew Eve and Christopher Burke, eds), *From Hieroglypics to Isotype* (London: Hyphen Press, 2010), p. 3.

61. E. H. Gombrich, *A Little History of the World* (first published in German, 1935; first edition in English, New Haven, CT and London: Yale University Press, 2005), p. 283.

62. Christoph Hermann, *Capitalism and the Political Economy of Work Time* (Abingdon: Routledge, 2015), pp 125–6.

63. https://thebeautyoftransport.wordpress.com/2013/11/06/glaze-amaze-tiled-railway-maps-of-the-uk.

64. Peter Barber and Tom Harper, *Magnificent Maps: Power, Propaganda and Art* (London: British Library, 2010), p. 11.

65. Henry Cole, *Travelling Charts*, published in *The Railway Chronicle* from 1846 onwards.

66. Francesca Carnevali and Julie Marie Strange, *20th Century Britain: Economic, Cultural and Social Change* (London: Routledge, 2007), p. 112.

67. Ron Ramdin, *Turning Pages,* vol. 1 (London: Compass, 2015), p. 212 and correspondence.

68. *Morden and Lea's Plan of London, 1682, also known as Ogilby and Morgan's Plan* (London: London Topographical Society, 1904).

69. Sir Herbert George Fordham, *Studies in Carto-Bibliography* (Oxford: Clarendon Press, 1914).

70. *Encyclopaedia Britannica*, 11th edn (New York: Encyclopaedia Britannica Company, 1911), vol. 17, p. 629.

71. Arthur Conan Doyle, *The Hound of the Baskervilles* (London: George Newnes, 1902), pp 52–3.

72. Erskine Childers, *The Riddle of the Sands: A Record of Secret Service Recently Achieved* (London: Smith, Elder & Co., 1903).

73. Pinda Bryars, unpublished research on crime maps.

74. Dorothy L. Sayers, *The Five Red Herrings* (London: Victor Gollancz, 1931) and *Gaudy Night* (London: Victor Gollancz, 1935).

75. Bryars and Harper, *History,* p. 132.

76. John Buchan, *The Thirty-Nine Steps* (Edinburgh and London: William Blackwood & Sons 1915), pp 42, 103 and 219.

77. Tom Conley, 'The 39 Steps and the Mental Map of Classical Cinema', in: Martin Dodge, Rob Kitchin and Chris Perkins (eds), *Rethinking Maps: New Frontiers in Cartographic Theory* (London: Routledge, 2009).

78. Mark Monmonier, *Air Apparent: How Meteorologists Learned to Map, Predict and Dramatize Weather* (Chicago: University of Chicago Press, 1999), p. 179.

79. Bryars and Harper, *History,* p. 202.

80. J. Brian Harley, *Ordnance Survey Maps: A Descriptive Manual* (Southampton: Ordnance Survey, 1975), p. 2.

81. Bryars and Harper, *History,* p. 100.

82. Bryars and Harper, *History,* p. 178.

83. Bryars and Harper, *History,* p. 187.

84. Bryars and Harper, *History,* p. 191.

85. William Cartwright, Georg Gartner and Antje Lehn (eds), *Cartography and Art* (Berlin: Springer, 2009), p. 12.

86. Gillian Gilbert *et al.*, *Bird Monitoring Methods: A Manual of Techniques for Key UK Species* (Sandy, Beds.: RSPB in association with British Trust for Ornithology, 1998), p. 44.

Chapter 4. Maps and Money

1. For a summary of the Marxist literature on capitalism see Anthony Brewer, *Marxist Theories of Imperialism: A Critical Survey*, 2nd edn (London: Routledge, 1990).

2. Diane Dillon, 'Consuming Maps', in James R. Akerman and Robert W. Karrow, Jr (eds), *Maps: Finding Our Place in the World* (Chicago and London: University of Chicago Press, 2007), pp 289–343; Catherine Delano-Smith, 'The Map as Commodity', in David Woodward (ed.), *Plantejaments i Objectius d'Una Història Universal de la Cartografia. 11è Curs, 21, 22, 23, 24i 25 de Febrer de 2000* ('Approaches and Challenges in a Worldwide History of Cartography') (Barcelona: Institut Cartogràfic de Catalunya, 2001), pp 91–109; Chandra Mukerji, *From Graven Images: Patterns of Modern Materialism* (New York and Guildford: Columbia University Press, 1983).

3. See Jeremy Black's chapter in this volume (chapter 1).

4. For an elegant summary of remote-sensing satellites see Laura Kurgan, *Close Up at a Distance: Mapping, Technology and Politics* (New York: Zone Books, 2013), pp 43–51.

5. The seminal work is David Woodward (ed.), *Five Centuries of Map Printing* (Chicago and London: University of Chicago Press, 1975). For the transition to photolithographic map production see Karen Severud Cook, 'The Historical Role of Photomechanical Techniques in Map Production', *Cartography and Geographic Information Science*, vol. 29, no. 3 (2002).

6. Mark Monmonier (ed.), The *History of Cartography, vol. 6: Cartography in the Twentieth Century* (Chicago: University of Chicago Press, 2015), p. 506.

7. For example, see Sarah Tyacke, *London Map-Sellers, 1660–1720: A Collection of Advertisements for Maps Placed in the 'London Gazette' 1668–1719, with Biographical Notes on the Map-Sellers* (Tring: Map Collector Publications, 1978).

8. Figures taken from the COPAC union catalogue of UK research libraries, http://copac.jisc.ac.uk/ [accessed 26 July 2016].

9. http://www.cartography.org.uk/downloads/MCT_BartsMaps.pdf [accessed 26 July 2016] – currency converted using http://www.nationalarchives.gov.uk/currency/results.asp#mid [accessed 26 July 2016]. For a sense of perspective, Thomas Jefferys's copper-engraved sheet map of the city of Louisburg in 1757 cost 2d, the equivalent of £8.52 in twenty-first century money.

10. George Philip & Son Ltd., Analysis Cost Books: customer orders C.8, C.20.

11. See, for example, *Bhutan: Landcover, Soil and Water Reflections from Landsat Imagery* (Washington DC: World Bank, 1982).

12. Eric Hobsbawm, *Age of Extremes: The Short Twentieth Century, 1914–1991* (London: Abacus, 1994), p. 289.

13. Jeffry A. Frieden, *Global Capitalism: Its Fall and Rise in the Twentieth Century* (New York and London: W. W. Norton, 2006), p. 168.

14. Paul R. Krugman, *Development, Geography and Economic Theory* (Cambridge, MA: MIT Press, 1995), p. 57.

15. Michael R. Curry, 'GIS and the Inevitability of Ethical Inconsistency', in John Pickles (ed.), *Ground Force: The Social Implications of Geographic Information Systems* (New York: Guildford Press, 1995), pp 68–87.

16. *History of Cartography*, vol. 6, p. 493, which comments on the discrepancies between the estimates.

17. Tony Judt, *Postwar: A History of Europe since 1945* (London: Vintage Books, 2010), p. 335.

18. *A New and Correct Map of the World Projected upon the Plane of the Horizon Laid Down from the Newest Discoveries and Most Exact Observations* (London: C. Price, sold by G. Willdey, 1714).

19. Peter Barber (ed.), *The Map Book* (London: Weidenfeld & Nicolson, 2005), pp 318–19.

20. Geoff King, *Mapping Reality: An Exploration of Cultural Cartographies* (Basingstoke: Macmillan, 1996).

21. Tim Bryars and Tom Harper, *A History of the 20th Century in 100 Maps* (London: British Library, 2014), pp 166–7.

22. Juergen Schultz, 'Jacopo di' Barbari's View of Venice: Map-Making, City Views and Moralized Geography Before the Year 1500', *Art Bulletin*, 60 (1978), pp 425–74.

23. Bryars and Harper, *History*, pp 140–1.

24. See Michael Heffernan's chapter in this volume (chapter 2) for further discussion of the United Nations logo.

25. Peter Barber and Tom Harper, *Magnificent Maps: Power, Propaganda and Art* (London: British Library, 2010), pp 166–7.

26. John R. Hébert, 'The Map that Named America', *Library of Congress Information Bulletin*, September 2003. https://www.loc.gov/loc/lcib/0309/maps.html [accessed 26 July 2016].

27. Bryars and Harper, *History*, pp 214–5

28. Hobsbawm, *Age of Extremes*, p. 90.

29. For the ethos behind this teaching see Halford Mackinder, 'The Teaching of Geography from an Imperial Point of View and the Use which Could and Should be Made of Visual Instruction', *Geographical Teacher*, 6 (1911), pp 79–86.

30. However, see Alexander Radó's 1936 *The Atlas of To-day and To-morrow*, mentioned in Tim Bryars's chapter in this volume (chapter 3).

31. J. A. Morris, quoted in Jeremy W. Crampton, 'Reflections on Arno Peters (1916–2002), *Cartographic Journal*, vol. 40, no. 1 (June 2003), p. 55.

32. Jeremy Black, *Maps and Politics* (London: Reaktion Books, 1996), pp 98–9.

33. King, *Mapping Reality*, p. 178–9.

34. Hobsbawm, *Age of Extremes*, p. 280.

35. Alastair Bonnett, *Off the Map: Lost Spaces, Invisible Cities, Forgotten Islands, Feral Places, and What They Tell Us about the World* (London: Aurum Press, 2015).

36. For these images see *England: The Photographic Atlas* (London: HarperCollins, 2001).

37. Judt, *Postwar*, p. 770.

38. See, for example, Siim Kallas, 'The Sea of Europe: Routing the Map for Economic Growth', 17 February 2014, http://europa.eu/rapid/press-release_SPEECH-14-133_en.htm [accessed 26 July 2016].

Chapter 5. Movement: Mapping Mobility to Mobile Mapping

1. David Harvey, *The Condition of Postmodernity: An Enquiry into the Origins of Cultural Change* (Oxford: Blackwell, 1990), p. 249.

2. Karl Marx and Friedrich Engels, *The Communist Manifesto*, introduction and notes by Gareth Stedman Jones (London: Penguin, 2002), p. 223.

3. Stephen Kern, *The Culture of Time and Space 1880–1918*, 2nd edn (Cambridge, MA: Harvard University Press, 2003).

4. Gilles Deleuze and Felix Guattari, *A Thousand Plateaus: Capitalism and Schizophrenia*, trans. Brian Massumi (Minneapolis: University of Minnesota Press, 1987), p. 385.

5. See Denis Cosgrove, 'Contested Global Visions: One-World, Whole-Earth, and the Apollo Space Photographs', *Annals of the Association of American Geographers*, vol. 84, no. 2 (1994), pp 270–94.

6. Mike Parker, *Mapping the Roads: Building Modern Britain* (Basingstoke: AA Publishing, 2014), p. 143.

7. Phyllis Pearsall, *From Bedsitter to Household Name: The Personal Story of A–Z Maps* (London: Geographers' A–Z Map Co., 1990). See also Simon Garfield, *On the Map: Why the World Looks the Way It Does* (London: Profile Books, 2012), pp 279–90.

8. Ken Garland, *Mr Beck's Underground Map* (Harrow Weald, Middx: Capital Transport Publishing, 1994).

9. Katherine Woollett and Eleanor A. Maguire, 'Acquiring "the Knowledge" of London's Layout Drives Structural Brain Changes', *Current Biology*, vol. 21, issue 24 (20 December 2011), pp 2109–14: dx.doi.org/10.1016/j.cub.2011.11.018 [accessed 26 July 2016].

10. For discussion of Berann's techniques, see Tom Patterson, 'A View from on High: Heinrich Berann's Panoramas and Landscape Visualization Techniques for the U.S. National Park Service', *Cartographic Perspectives*, no. 36 (Spring 2000), pp 38–65. All his work can be seen at www.berann.com [accessed 26 July 2016].

11. Wayne G. Hammond and Christina Scull, *The Art of 'The Lord of the Rings' by J.R.R. Tolkien* (London: HarperCollins, 2015), p. 12.

12. Robert A. Stallings, 'Evacuation Behavior at Three Mile Island', *International Journal of Mass Emergencies and Disasters*, vol. 2, no. 1 (1984), pp 11–26.

13. For discussions of this problem and cartographers' strategies to resolve it, see Irina Ren Vasiliev, 'Mapping Time', *Cartographica*, vol. 34, no. 2 (Summer 1997), pp 1–51, and Daniel Rosenberg and Anthony Grafton, *Cartographies of Time* (New York: Princeton Architectural Press, 2010).

14. See the artist's biography at Stefan R. Landsberger's website, 'Chinese Posters: Propaganda, Politics, History, Art', hosted by the International Institute of Social History, Amsterdam. chineseposters.net/artists/liuwenxi.php [accessed 26 July 2016].

15. See Arthur H. Robinson, *Early Thematic Mapping in the History of Cartography* (Chicago, University of Chicago Press, 1982).

16. Dieter Roelstraete, *Richard Long: A Line Made by Walking* (London: Afterall, 2010).

17. Kate Brown, *Biography of No Place: From Ethnic Borderland to Soviet Heartland* (Cambridge, MA: Harvard University Press, 2004), p. 209; see also pp 197–8.

18. For an outline of the COBE mission, see www.science.nasa.gov/missions/cobe/ [accessed 26 July 2016].

19. All of Berann's ocean floor maps can be seen at www.berann.com/panorama/archive/index.html#-Ocean_Floor_Maps [accessed 26 July 2016].

20. The OS website gives details of its revision policy at www.ordnancesurvey.co.uk/about/governance/policies/basic-scale-revision.html [accessed 26 July 2016].

21. Jean M. Grove, *The Little Ice Age*, 2nd edn, vol. 1 (London: Routledge, 2004), pp 52–3.

22. See David J. A. Evans, 'Glacial Geomorphology at Glasgow', *Scottish Geographical Journal*, vol. 125, nos. 3–4 (2009), pp 285–320. dx.doi.org/10.1080/14702540903364310 [accessed 26 July 2016].

23. Rob McKie, 'Now Europe's Biggest Glacier Falls to Global Warming', *Guardian*, 22 October 2000. www.theguardian.com/environment/2000/oct/22/weather.climatechange [accessed 10 March 2016].

24. Editorial, 'El Chamizal Dispute Between the United States and Mexico', *American Journal of International Law*, vol. 4, no. 4 (October 1910), pp 925–30. archive.org/details/jstor-2186807 [accessed 26 July 2016].

25. Paul Kramer, 'A Border Crosses', *New Yorker*, 20 September 2014. www.newyorker.com/news/news-desk/moving-mexican-border [accessed 26 July 2016].

26. Mark Monmonier, 'Strategies for the Vizualization of Geographic Time-Series Data', *Cartographica*, vol. 27, no. 1 (Spring 1990), pp 30–45.

27. An animation of the postcard can be viewed at labs.wolfsonian.org/animation/img/Hungary1918-treatyDivision.gif [accessed 26 July 2016].

28. Sébastien Caquard and D. R. Fraser Taylor, 'What is Cinematic Cartography?', *Cartographic Journal*, vol. 46, no. 1 (2009), pp 5–8.

29. For an extended analysis, see Tom Conley, *Cartographic Cinema* (Minneapolis: University of Minnesota Press, 2007), pp 93–100.

30. Sébastien Caquard, 'Foreshadowing Contemporary Digital Cartography: A Historical Review of Cinematic Maps in Films', *Cartographic Journal*, vol. 46, no. 1 (2009), pp 46–55; Mark Dorrian, 'On Google Earth', in Mark Dorrian and Frédéric Pousin (eds), *Seeing from Above: The Aerial View in Visual Culture* (London: I. B. Tauris, 2013), pp 290–307.

31. Norman Thrower, 'Animated Cartography', *Professional Geographer*, vol. 11, no. 6 (1959), pp 9–12.

32. Val Noronha, 'In-Vehicle Navigation System', in Mark Monmonier (ed.), *The History of Cartography: Volume Six: Cartography in the Twentieth Century* (Chicago: University of Chicago Press, 2015), pp 1716–22.

33. Ralph E. Ehrenburg, 'Aeronautical Chart', in Monmonier (ed.), *The History of Cartography*, pp 28–9.

34. Avidyne advertisement, *Flying*, March 1999, p. 94.

35. See the essays in Dorrian and Pousin, *Seeing from Above*.

36. For example, Peter Adey, *Aerial Life: Spaces, Mobilities, Affects* (Chichester: Wiley-Blackwell, 2010);

Laura Kurgan, *Close Up at a Distance: Mapping, Technology and Politics* (New York: Zone Books, 2013).

37. Peter Collier, 'The Impact on Topographic Mapping of Developments in Land and Air Survey: 1900–1939', *Cartography and Geographic Information Science*, vol. 29, no. 3 (2002), pp 155–74.

38. Kurgan provides a succinct overview of the evolution of satellite mapping in *Close Up at a Distance*, pp 39–54. See also Stephen S. Hall, *Mapping the Next Millennium. How Computer-Driven Cartography is Revolutionizing the Face of Science* (New York: Vintage Books, 1992), pp 52–138.

39. See Kenneth Cox, Stephen G. Eick and Taosong He, '3D Geographic Network Displays', ACM *SIGMOD Record*, vol. 25, no. 4 (1996), pp 50–4: dl.acm.org/citation.cfm?id=245901 [accessed 26 July 2016].

40. See Emmanuel Frécon, 'WebPath – a 3D Browsing History Visualisation', *ERCIM News*, no. 41 (2000). www.ercim.eu/publication/Ercim_News/enw41/frecon.html [accessed 26 July 2016]; Martin Dodge and Rob Kitchen, *Mapping Cyberspace* (London: Routledge, 2003), pp 126–7.

41. William J. Rankin, 'Global Positioning System', in Monmonier (ed.), *History of Cartography*, pp 551–8; Bernhard Hofmann-Wellenhof, Herbert Lichtenegger and Elmar Wasle, *GNSS – Global Navigation Satellite Systems: GPS, GLONASS, Galileo, and More* (Vienna: Springer, 2008).

42. Peter Galison, *Einstein's Clocks: Poincaré's Maps* (New York: W. W. Norton, 2003), p. 285.

43. Rankin, 'Global Positioning System', p. 555.

44. Harvey, *Condition of Postmodernity*.

45. Gilles Deleuze and Felix Guattari, *A Thousand Plateaus*.

46. Jerome E. Dobson and Peter F. Fisher, 'The Panopticon's Changing Geography', *Geographical Review*, vol. 97 (2007), pp 307–23. See also Kurgan, *Close Up at a Distance*, pp 9–36. For a counter-argument that Google Earth is an empowering and emancipating rather than imperialist project, see Jason Farman, 'Mapping the Digital Empire: Google Earth and the Process of Postmodern Cartography', *New Media and Society*, vol. 12, no. 6 (September 2010), pp 869–88. www.inter-disciplinary.net/ci/cyber%20hub/visions/v3/Farman%20paper.pdf [accessed 26 July 2016].

47. Rebecca Solnit, *A Field Guide to Getting Lost* (Edinburgh: Canongate, 2005).

48. Garfield, *On the Map*, pp 372–84; Mike Parker, *Map Addict* (London: Collins, 2010), pp 277–93.

49. John Pickles, 'Preface', in Pickles (ed.), *Ground Truth: The Social Implications of Geographic Information Systems* (New York: Guilford Press, 1995), pp x, xii.

50. Surveillance Studies Network, 'A Report on the Surveillance Society for the Information Commissioner', September 2006. www.surveillance-studies.net/?page_id=3 [accessed 26 July 2016]. The SSN website uses Google Maps to plot the location worldwide of researchers and research projects in surveillance studies. For legal perspectives, see 'Privacy and Big Data: Making Ends Meet', *Stanford Law Review Online: Symposium Issue*, 3 September 2013. www.stanfordlawreview.org/online/privacy-and-big-data [accessed 26 July 2016].

51. For OpenStreetMap see www.openstreetmap.org/about [accessed 26 July 2016] and for an example of its deployment see the Humanitarian OpenStreetMap Team, hotosm.org [accessed 26 July 2016]. There is a large and growing scholarship in 'critical GIS'. For an overview, see Jeremy W. Crampton, *Mapping: A Critical Introduction to Cartography and GIS* (Chichester: Wiley-Blackwell, 2010).

BIBLIOGRAPHY

James R. Akerman (ed.), *The Imperial Map: Cartography and the Mastery of Empire* (Chicago: University of Chicago Press, 2009)

Peter Barber (ed.), *The Map Book* (London: Weidenfeld & Nicolson, 2005)

Peter Barber and Tom Harper, *Magnificent Maps: Power, Propaganda and Art* (London: The British Library, 2010)

Timothy Barney, *Mapping the Cold War: Cartography and the Framing of America's International Power* (Chapel Hill, NC: The University of North Carolina Press, 2015)

Jeremy Black, *Maps and Politics* (London: Reaktion Books, 1996)

Alastair Bonnett, *Off the Map: Lost Spaces, Invisible Cities, Forgotten Islands, Feral Places, and What they Tell Us About the World* (London: Aurum Press, 2015)

Tim Bryars and Tom Harper, *A History of the Twentieth Century in 100 Maps* (Chicago: University of Chicago Press, 2014)

Francesca Carnevali and Julie Marie Strange, *20th Century Britain: Economic, Cultural and Social Change* (London: Routledge, 2007)

William Cartwright, Georg Gartner, Antje Lehn (eds.), *Cartography and Art* (Berlin: Springer, 2009)

Peter Chassaud, *Mapping the First World War* (Glasgow: Collins, 2013)

J. J. Cherns, *Official Publishing: An Overview* (Oxford: Pergamon Press, 2013)

Peter Collier, 'The Impact on Topographic Mapping of Developments in Land and Air Survey: 1900–1939', *Cartography and Geographic Information Science*, vol. 29, no. 3 (2002)

Tom Conley, *Cartographic Cinema* (Minneapolis: University of Minnesota Press, 2007)

Denis Cosgrove, *Apollo's Eye: A Cartographic Genealogy of the Earth in the Western Imagination* (Baltimore, MD: Johns Hopkins University Press, 2001)

Jeremy W. Crampton, *Mapping: A Critical Introduction to Cartography and GIS* (Chichester: Wiley-Blackwell, 2010)

Jeremy W. Crampton, 'Cartography's Defining Moment: the Peters Projection Controversy, 1974–1990', *Cartographica*, vol. 31, no. 4 (1994)

Diane Dillon, 'Consuming Maps' in James R. Akerman and Robert W. Karrow Jr (eds.), *Maps: Finding Our Place in the World* (Chicago: University of Chicago Press, 2007)

Martin Dodge and Rob Kitchen, *Mapping Cyberspace* (London: Routledge, 2003)

Martin Dodge, Rob Kitchin and Chris Perkins (eds), *Rethinking Maps: New Frontiers in Cartographic Theory* (London: Routledge, 2009

Catherine Dunlop, *Cartophilia: Maps and the Search for Identity in the French-German Borderland* (Chicago: University of Chicago Press, 2015)

Stuart Elden, *The Birth of Territory* (Chicago: University of Chicago Press, 2013)

Jason Farman, 'Mapping the Digital Empire: Google Earth and the Process of Postmodern Cartography', *New Media and Society*, vol. 12, no. 6 (September 2010)

Peter Galison, *Einstein's Clocks: Poincaré's Maps* (New York: W. W. Norton, 2003)

Simon Garfield, *On the Map. Why the World Looks the Way it Does* (London: Profile Books, 2012)

Ken Garland, *Mr Beck's Underground Map* (Harrow Weald: Capital Transport Publishing, 1994)

Stephen S. Hall, *Mapping the Next Millennium. How Computer-Driven Cartography is Revolutionizing the Face of Science* (New York: Vintage Books, 1992)

Wayne G. Hammond and Christina Scull, *The Art of 'The Lord of the Rings' by J.R.R. Tolkien* (London: HarperCollins, 2015)

J. B. Harley, *Ordnance Survey Maps: A Descriptive Manual* (Southampton: Ordnance Survey, 1975)

David Harvey, *The Condition of Postmodernity: An Enquiry into the Origins of Cultural Change* (Oxford: Blackwell, 1990)

Michael Heffernan, *The Cartography of the Fourth Estate: Mapping the New Imperialism in British and French Newspapers c. 1875–1925*, in James Akerman (ed.), *The Imperial Map: Cartography and the Mastery of Empire* (Chicago: University of Chicago Press 2009)

Eric Hobsbawm, *Age of Extremes: The Short Twentieth Century, 1914–1991* (London: Abacus, 1994)

Tony Judt, Postwar: *A History of Europe since 1945* (London: Vintage, 2010)

Stephen Kern, *The Culture of Time and Space 1880–1918*, 2nd edn (Cambridge, MA: Harvard University Press, 2003)

Geoff King, *Mapping Reality: An Exploration of Cultural Cartographies* (Basingstoke: Macmillan, 1996)

Laura Kurgan, *Close up at a Distance: Mapping, Technology and Politics* (New York: Zone Books, 2013)

Mark Monmonier, *The History of Cartography, vol. 6: Cartography in the Twentieth Century* (Chicago: University of Chicago Press, 2015)

Mark Monmonier, *Rhumb Lines and Map Wars: A Social History of the Mercator Projection* (Chicago: University of Chicago Press, 2004)

Alastair Pearson, D. R. Fraser Taylor, Karen Kline and M. Heffernan, 'Cartographic Ideals and Geopolitical Realities: International Maps of the World from the 1890s to the Present', *Canadian Geographer*, vol. 50, no. 2 (2006)

Mike Parker, *Mapping the Roads: Building Modern Britain* (Basingstoke: AA Publishing, 2014)

Mike Parker, *Map Addict* (London: Collins, 2010), pp 277–93

Phyllis Pearsall, *From Bedsitter to Household Name: The Personal Story of A-Z Maps* (London: Geographers' A–Z Map Company, 1990)

Alastair Pearson and Michael Heffernan, 'The American Geographical Society's Map of Hispanic America: Million Scale Mapping Between the Wars', Imago Mundi, vol. 6, no. 2 (2009)

John Pickles, A History of Spaces: Cartographic Reason, Mapping, and the Geo-coded World (London: Routledge, 2004)

John Pickles (ed.), *Ground Force: The Social Implications of Geographic Information Systems* (New York: Guildford Press, 1995)

Arthur H. Robinson, *Early Thematic Mapping in the History of Cartography* (Chicago: University of Chicago Press, 1982)

Daniel Rosenberg and Anthony Grafton, *Cartographies of Time* (New York: Princeton Architectural Press, 2010)

Susan Schulten, *The Geographical Imagination in America 1880–1945* (Chicago: University of Chicago Press, 2001)

Norman Thrower, 'Animated Cartography', *Professional Geographer*, vol. 11, no. 6 (1959)

David Woodward (ed.), *Five Centuries of Map Printing* (Chicago: University of Chicago Press, 1975)

ILLUSTRATION CREDITS

Grateful thanks to all individuals and institutions who assisted with the images included in this book. Unless otherwise specified, images are © The British Library Board. Credits listed below are by image number.

2 © Alighiero Boetti by SIAE/DACS 2016
3 Topical Press Agency/Getty Images
9 © BBC Photo Library
16 © Estate of Guy Debord
31 FDR Presidential Library & Museum
33, 34 Persuasive Maps: The PJ Mode Collection
35 Keystone/Getty Images
39 123rf
42, 43 League of Nations Photo Archive
46 Michael Ochs Archive/Getty Images
47 © Giuseppe/Panoramio
48 © BBC Photo Library
58 www.peacenow.org
69 National Railway Museum/SSPL
71 © The Estate of Edward Bawden / British Airways
77 Gaumont-British/The Kobal Collection
102 © The Estate of Macdonald Gill
106 British Cartoon Archive, © Ralph Steadman
111 K-Photos/Alamy Stock Photo
114, 115 © TfL from the London Transport Museum Collection
116 Carl Court/Getty Images
120 © The Estate of J.R.R. Tolkien
125, 126 © Richard Long. All rights reserved, DACS 2016
138 © Jeremy Wood

ACKNOWLEDGEMENTS

It has taken the skill and generosity of many people to make this exhibition and book happen. In particular I wish to thank Tim Bryars, whose ideas and research have focused greater attention on twentieth-century maps in recent years. The project was expertly advised by Nick Baron and Mike Heffernan of the University of Nottingham, and I thank them for many insightful conversations. The interest and support of Tony and Maureen Wheeler, who played such an important role themselves in the cartographic story of the twentieth century, has been invaluable. Since his retirement as the British Library's Map Librarian in September 2015, Peter Barber has been a constant source of wisdom and encouragement. The perceptive observer will have noted his methodology running through the exhibition. I am most grateful to the Friends of the British Library for their kind support.

The following also deserve my thanks: Alastair Bonnett, Reginald and Caroline Byron of Tangmere Military Aviation Museum, Jeremy Crampton, Catherine Delano-Smith, Mark Durston, Stephen Eick, Stuart Elden, Kenneth Field, Tom Harper of Avidyne, Kat Hubschmann of the Wiener Library, Ed King, Roland-Francois Lack, P.J. Mode, Miran Norderland, Ed Parsons, Rob Sharpe, Chloe Smith from Gamecity, Fabian Tompsett, David Welch, William Wells, Shaun White of Electronic Arts, and Jeremy Wood. The exhibition design was provided by Mel Northover and Ali Brown of Northover Brown: thank you for turning the concept into reality.

I would like to thank the following people who were instrumental in enabling works to be borrowed for the exhibition. Caterina Raganelli Boetti and Dr Francesca Franco of the Fondazione Aligiero e Boetti; Christopher Tolkien (who once kindly responded to a ten year-old's fan mail), the Tolkien Family Trust, Cathleen Blackburn, and Catherine Parker of Bodleian Libraries; Adrian Seville; Gerry Connolly of Worthing Museum and Art Gallery; Christopher Marsden and Sara Mittica of the Victoria & Albert Museum; Andrew Wilson and Sanne Klinge of Tate, Juergen Vervoorst and Kate Narewska from The National Archives at Kew, and Taz Chappell and Tim Clark of the British Museum.

This project has only been possible because of the dedication and professionalism of colleagues at the British Library. Particularly, I wish to thank the perennially unsung heroes of the Library's Exhibitions team Janet Benoy, Susan Dymond and Alex Kavanagh. Thank you to Robert Davies, Sally Nicholls and Miranda Harrison of British Library Publishing, and to Briony Hartley of Goldust Design, who managed this publication with customary patience and flexibility.

Thanks are also due to Jamie Andrews, who first convinced me to curate a twentieth-century map exhibition; The Registrars' office led by Barbara O'Connor; Head of Conservation Cordelia Rogerson with Liz Rose, Mark Browne and his team; Silvia Dobrovich and Alex Michaels of the Development Office; The Press team including Sophie McIvor, Elsie King and Ben Sanderson; John Overeem of the Design Office; Tony Grant from Imaging Services, and Phil Davies and Fran Fuentes of the Eccles Centre for American Studies.

I'm grateful for the invaluable advice, honesty and support of curatorial colleagues, especially Alison Bailey, Jim Caruth, Andrea Clarke, Ian Cooke, Nick Dykes, Rachel Foss, Philip Hatfield, Julian Harrison, Crispin Jewitt, Helen Melody, Chris Michaelides, Hamish Todd, Helen Peden, Magdalena Peszko, Barry Taylor, Marion Wallace, Zoe Wilcox and Stella Wisdom. The staff of the Map Library have given substantial support and my thanks go to Nicola Beech, Carols Garcia, Sue Young, Kate Marshall and Magda Kowalzcuk, as well as Jacqueline Pitcher, Janet Grover and their respective teams.

I would like to thank my wife Chloé for allowing me the demanding hours needed to make this happen, and for keeping me in the real world throughout. Finally, thank you to everyone whose personal recollections of the twentieth century have breathed life into the exhibition and the pages of this book.

INDEX

Figures in *italic* refer to pages on which
illustrations appear